# Applied Statistics
# and the SAS Programming Language

Second Edition

# Applied Statistics and the SAS Programming Language

## Second Edition

### Ronald P. Cody
University of Medicine and Dentistry of New Jersey
Robert Wood Johnson Medical School

### Jeffrey K. Smith
Graduate School of Education
Rutgers University

North-Holland
New York • Amsterdam • London

Note
SAS is a registered trademark of SAS Institute Inc., Cary, North Carolina.

Elsevier Science Publishing Co., Inc.
52 Vanderbilt Avenue, New York, New York 10017

Sole distributors outside the U.S. and Canada:
Elsevier Science Publishers B.V.
P.O. Box 211, 1000 AE Amsterdam, The Netherlands

Current printing (last digit):
10   9   8   7   6   5   4   3   2   1

Manufactured in the United States of America

519.5
C673

# Contents

153/139

# Preface to the Second Edition

You may ask, "Why publish a second edition so soon after the first?" There have been several **major** changes to the SAS® system since the publication of the first edition of this book in early 1985. First, the SAS Institute released version 5 for mainframe computers in 1985. Several important changes were made, both in syntax and statistical capability. For example, our chapter on repeated measures analysis of variance had to be rewritten to accommodate the new **REPEATED** statement of PROC GLM and ANOVA.

Probably the biggest change to SAS software was the introduction of **SAS/PC**® in the Fall of 1985. A chapter devoted entirely to SAS/PC is included in this edition. Changes and additions to all chapters of the book were also necessary.

A **reference section** was added to our book. The first reference chapter is the one mentioned above, devoted to SAS/PC. The second reference chapter contains the solution to common SAS and data analysis questions. The third reference chapter is for advanced SAS users and contains many tricks and advanced uses of SAS software by way of two examples. Finally, we expanded the problem sets by over 50% so that the book would be more usable as a text.

As the authors and others used the book over the past two years, errors were identified. We hope these have been corrected in this edition and that we have not made too many new ones!

We hope that you will find this new edition useful and enjoyable. Happy computing.

# Preface to the First Edition

Most researchers are inherently more interested in the substance of their research endeavors than in statistical analyses or computer programming. Yet, conducting these analyses is an integral link in the research chain. All too frequently, it is the weak link. This is particularly true when there is no resource for the applied researcher to look to for assistance in running computer programs for statistical analyses. **Applied Statistics and the SAS Programming Language** is intended to provide the applied researcher with the capacity to perform statistical analyses with SAS® software without wading through pages of technical documentation.

The researcher is provided with the necessary SAS statements to run programs for most of the commonly used statistics, explanations of the computer output, interpretations of results, and examples of how to construct tables and write-up results for reports and journal articles. Examples have been selected from business, medicine, education, psychology, and other disciplines.

We would like to acknowledge the contributions of Alissa Bernholc and Michele Demak, who spent many hours reading and editing this text. Many errors of omission and commission were caught by their careful attention to detail.

SAS is a registered trademark of SAS Institute Inc., Cary, N.C.

# Applied Statistics
# and the SAS Programming Language

Second Edition

# Chapter 1    Getting Started

## A. THE SCOPE AND PURPOSE OF THIS BOOK

This book is about using the computer for the statistical analysis of data. It is a "how-to" book rather than a theoretical discussion of statistics. To this end, the focus of the book is the practical application of a computer package called SAS in the analysis of data sets. (SAS is a registered trademark of the SAS Institute, Cary, NC.) The intended reader of this text is the applied researcher in any one of a variety of fields: medicine, business, education, the social sciences, etc. Our goal is to provide a comprehensible guide to the analysis and interpretation of data from a nontechnical perspective. We have decided to base this guide on SAS software for the following reasons.

(1) SAS software is currently available on hundreds of computer facilities and the number is increasing.

(2) SAS software is now available on IBM-compatible microcomputers.

(3) SAS programming is easy to use.

(4) SAS software allows a variety of sophisticated as well as elementary analyses and has tremendous flexibility.

SAS software has routines for describing data and generating statistical analyses. In order to analyze our data, we will need to write SAS program statements that describe our data and request statistical calculations. On mainframe computers, depending on the operating system, you may need to add control information to identify yourself to the computer and to tell the computer that you are running a SAS program (as compared to COBOL or FORTRAN, for example). This information is either contained in what is called "job control language" or is controlled by the operating system (CMS or TSO, for example). You may be punching the control cards, the SAS instructions, and your data on cards with a keypunch or you may be typing the same information into the computer via a terminal. If you are using

SAS/PC (SAS system on a microcomputer), you will be interacting with the SAS Display Manager. As the control information is different from one mainframe computer to the next, you will need to seek assistance from someone at your local computer center. Users of SAS/PC will find a special chapter devoted to running SAS software on microcomputers (Reference Chapter 1).

One of the advantages of SAS software is the variety of procedures that can be performed. We cannot describe them all here. It is therefore recommended that the following SAS manuals be obtained:

**Mainframe Manuals**

Introductory Guide, Version 5 Edition
SAS User's Guide:Basics, Version 5 Edition
SAS User's Guide:Statistics, Version 5 Edition

**Microcomputer-Based Manuals**

SAS Introductory Guide for Personal Computers
SAS Language Guide for Personal Computers
SAS Procedures Guide for Personal Computers
SAS/STAT Guide, for Personal Computers

These manuals may all be obtained from

SAS Institute Inc.
Box 8000
Cary, North Carolina 27511-8000
Telephone (919) 467-8000, ext. 7001

The statistics discussed in this book vary greatly in terms of complexity. On one hand we have a discussion of means and standard deviations, on the other, repeated measures analysis of variance and multiple regression. Again let us emphasize that many of the details of the statistical procedures presented here will be sufficient only to understand (1) when a particular procedure is to be used, and (2) how to interpret the results. A good statistics text should accompany this book. We recommend

<u>Statistical Principles in Experimental Design</u>, by B.J. Winer (McGraw-Hill, New York, 1971)

<u>Statistical Methods</u>, by Snedecor and Cochran (Iowa State University Press, Iowa, 1980)

<u>Multiple Regression in Behavioral Research:</u> <u>Explanation and Prediction</u>, by Elazar J. Pedhazur (Holt, Rinehart and Winston, New York, 1982)

<u>Experimental Design in Psychological Research</u>, by Edwards (Harper & Row, New York, 1985)

<u>Multivariate Statistics in Behavioral Research</u>, by R. Darrell Bock (McGraw-Hill, New York, 1975)

## B. INTRODUCTION

Computers are like Martians. They are simultaneously highly intelligent and remarkably ignorant. They can perform wonderfully complex manipulations but only if instructed in a precise, detailed fashion. This is similar to the way Hollywood depicted Martians in the movies we saw when we were kids. A Martian could tell you what day of the week October 12, 1874 was, but would not be insulted if you called him a turkey.

So it is with computers. They require detailed instructions on how to behave in order to get what we want. Since the computer only truly understands whether certain of its parts are electrically charged or not, it is necessary for computer scientists to provide a way for the average human to communicate with the computer. For those of us average humans who wish to do statistical analysis, SAS software is the "language" to use.

SAS software was designed primarily to perform statistical analyses and report the results. It also helps us store, arrange, and report on large amounts of data. In this section, we will focus on the rudiments of using SAS software, and the computer in general. Many people run SAS programs on IBM computers, either (1) as "batch" programs, which are run by reading cards into a card reader connected to the computer; (2) by entering the program using a terminal connected to the computer; (3) in an interactive mode where the SAS program is controlled in a more direct fashion (examples are versions of SAS software that run under CMS or TSO). With the introduction of SAS/PC, many people are using microcomputers to run SAS software. Don't let these descriptions of computer systems intimidate you. All that is necessary is to find out who at your business or university can provide you with the proper information concerning which computer system you will be using and what type of SAS installation you have. SAS programs will also run on a variety of minicomputers (not to be confused with your home "Martian Attack" video player). No matter which computer system you use, you will need to start any SAS program with some statements particular to that system. With a "batch" system, these statements typically include a "JOB" statement, which tells the computer, among other things, whom to bill for using the system, and a statement or two that tells the computer to run your SAS program. Since these statements vary from system to system, you must get this information from your local computer center.

## C. COMPUTING WITH SAS SOFTWARE

Assuming you have obtained such information, we can begin our discussion of SAS programming. SAS programs communicate with the computer by SAS "statements." There are several kinds of SAS statements, but they share a common feature--they end in a semicolon. **A semicolon in a SAS program is like a period in English.** Probably the most common error found in SAS programs is the omission of the semicolon. This causes the computer to read two statements as a run-on statement and invariably fouls things up.

There are two kinds of SAS statements: **DATA** statements and **PROC** statements. DATA statements tell SAS programs about your data. They are used to indicate where the variables are on data lines, what you want to call the variables, and how to create new variables from existing variables, as well as for several other functions we will mention later. PROC statements (pronounced "prock"; this is short for "PROCEDURE") indicate what kind of statistical analyses to perform and provide specifications for those analyses.

SAS programs are like sandwiches. They begin with DATA statements (bread), which tell the program about your data set. Then come the data (meat). Finally, we have PROC statements (bread), which specify the analyses to be performed. Let's look at an example. Consider this simple data set:

| SUBJECT NUMBER | SEX (M OR F) | EXAM 1 | EXAM 2 | HOMEWORK GRADE |
|---|---|---|---|---|
| 10 | M | 80 | 84 | A |
| 7  | M | 85 | 89 | A |
| 4  | F | 90 | 86 | B |
| 20 | M | 82 | 85 | B |
| 25 | F | 94 | 94 | A |
| 14 | F | 88 | 84 | C |

We have 5 variables (SUBJECT NUMBER, SEX, EXAM 1, EXAM 2, and HOMEWORK GRADE) collected on each of 6 subjects. The unit of analysis, subjects in this example, is called an **observation** in SAS terminology. SAS software uses the term **variable** to represent each piece of information we collect for each observation. Before we can write our SAS program, we need to assign a **variable name** to each variable. We do this so that we can distinguish one variable from another when doing computations or requesting statistics. SAS variable names must conform to a few simple rules: They must start with a letter, be no more than 8 characters in length, and not contain blanks or special characters such as commas, semicolons, etc. Therefore, our column headings of "SUBJECT NUMBER," or "EXAM 1" are not valid SAS variable names. Logical SAS variable names for this collection of data would be:

SUBJECT   SEX     EXAM1     EXAM2     HWGRADE

It is usually wise to pick variable names that help you remember which name goes with which variable. We could have named our 5 variables VAR1, VAR2, VAR3, VAR4, and VAR5, but we would then have to remember that VAR1 stands for "SUBJECT NUMBER," etc.

To begin, let's say we are interested only in getting the class means for the two exams. In reality it's hardly worth engaging several million dollars worth of machinery to add up some numbers, but it does provide a nice example. In order to do this, we could write the following SAS program:

```
1    DATA TEST;
2    INPUT SUBJECT 1-2 SEX $ 4 EXAM1 6-8 EXAM2 10-12
     HWGRADE $ 14;
3    CARDS;
4    10 M  80   84 A
5     7 M  85   89 A
6     4 F  90   86 B
7    20 M  82   85 B
8    25 F  94   94 A
9    14 F  88   84 C
10   PROC MEANS;
```

NOTE:   The line numbers to the left of the program  are
**not** part of the program.  They are just there so that we
can refer to them in the text.

   **Lines 1-3**  are  the  DATA  statements or DATA paragraph.   They begin with the word DATA and end  with the
word  CARDS.    Even though SAS programs use a  statement
called **"CARDS,"**   it does not mean that our data have to
be on physical cards. Rather, it refers to the more general  concept "card image," which can be real cards  or
records  on  a computer disk.  **Line 1** tells the  program
that we want to create a SAS data set called **TEST**.    The
data set name is optional and  follows the same rules as
a variable name.  **Line 2** is an **INPUT** statement and gives
the program two pieces of information:  what to call the
variables and where to find them on the data cards.   The
first variable is SUBJECT and can be found in columns  1
and 2 of the data cards.  The second variable is SEX and
can  be found in column 4 of the data cards. The  dollar
sign after SEX is **not** what you think.   It simply  means
that SEX is an  "alphanumeric"  variable, that  is,  a
variable can have letters as data values.  More on  this
later. EXAM1 is in columns 6-8 etc. **Line 3** says that the
DATA  statements are done and the next thing the program
should  look  for are the data  themselves.   **Lines  4-9**
contain our data.

   There  is  great  latitude  possible  in  putting
together  the  data cards.  Using a few rules will  make
life much simpler for you. These are not **laws**,  they are
just suggestions.  First,  put each new observation on a
new card.  Having more than one card per observation  is
often necessary (and no problem),  but don't put two ob-

servations on one card (at least for now). Second, line up your variables. Don't put EXAM1 in columns 6-8 on one card and in columns 7-8 on the next. SAS software can actually handle some degree of sloppiness here, but sooner or later it'll cost you. Third, right-justify your data values. If you have AGE as a variable, record the data as follows:

| CORRECT | PROBLEMATIC |
|---------|-------------|
| 87      | 87          |
| 42      | 42          |
| 9       | 9           |
| 26      | 26          |
| 4       | 4           |

**Right justified      Left justified**

Once again, SAS software doesn't care whether you right-justify or not, but other statistical programs will, and right justification is standard. Fourth, use alpha-numeric variables sparingly. Take HWGRADE for example. We have HWGRADE recorded alphanumerically. (Normal people call this alphabetically. Computer people don't because letters and/or numbers can be treated as alpha-numerics.) But we could have recorded it as 0-4 (0=F, 1=D, etc.). As it stands, we **cannot** compute a mean grade. Had we coded HWGRADE numerically, we could get an average grade. Enough on how to code data for now.

SAS programs know that the data cards are over when it finds a SAS statement. Usually, the next SAS state-ment is a PROC statement. PROC says, "Run a procedure," to the program. We specify which procedure right after the word PROC. **Line 10** says to run the procedure called MEANS. The MEANS procedure calculates the mean for any variables you specify.

When this program is executed, it produces some-thing called the **SAS LOG** and the SAS **OUTPUT.** The SAS LOG is an annotated copy of your original program (without the data listed). Any SAS error messages will be found there, along with information about the data set that was created. The SAS LOG for this program is shown below:

```
1 S A S   L O G     OS SAS 82.3   VS2/MVS JOB EX1  STEP SAS  PROC
                                   11:55 WEDNESDAY, JULY 23, 1986
NOTE: THE JOB EX1 HAS BEEN RUN UNDER RELEASE 82.3 OF SAS AT
   N. J. EDUCATIONAL COMPUTER NETWORK INC. (01014001).
NOTE: SAS OPTIONS SPECIFIED ARE:
   SORT=4

1           DATA TEST;
2           INPUT SUBJECT 1-2 SEX $ 4 EXAM1 6-8 EXAM2 10-12
3           HWGRADE $ 14;
4           CARDS;

NOTE: DATA SET WORK.TEST HAS 6 OBSERVATIONS AND 5 VARIABLES.
                                             635 OBS/TRK.
NOTE: THE DATA STATEMENT USED 0.04 SECONDS AND 72K.

11          PROC MEANS;

NOTE: THE PROCEDURE MEANS USED 0.13 SECONDS AND 142K AND
                                        PRINTED PAGE 1.
NOTE: SAS USED 142K MEMORY.
NOTE: SAS INSTITUTE INC.
   SAS CIRCLE
   PO BOX 8000
   CARY, N.C.  27511-8000
```

The  more important part of the output contains the results of the computations and procedures requested  by our PROC statements. This portion of the output from the above program is shown next:

```
SAS                  11:59 WEDNESDAY, JULY 23, 1986              1

VARIABLE  N     MEAN       STANDARD       MINIMUM       MAXIMUM
                           DEVIATION       VALUE         VALUE

SUBJECT   6  13.33333333  7.99166232   4.00000000   25.00000000
EXAM1     6  86.50000000  5.20576603  80.00000000   94.00000000
EXAM2     6  87.0000000   3.89871774  84.00000000   94.00000000


VARIABLE      STD ERROR       SUM        VARIANCE        C.V.
              OF MEAN

SUBJECT      3.26258248   80.00000000  63.86666667     59.937
EXAM1        2.12524508  519.00000000  27.10000000      6.018
EXAM2        1.59164485  522.00000000  15.20000000      4.481
```

If you don't specify which variables you want, SAS software will calculate the mean for every numeric variable in the data set. Our program calculated means for SUBJECT, EXAM1, and EXAM2. Since SUBJECT is just an arbitrary ID number assigned to each student, we aren't really interested in its mean. We can avoid getting it (and paying for it) by adding a new statement under PROC MEANS:

```
10      PROC MEANS;
11          VAR EXAM1 EXAM2;
```

The indentation is only a visual aid. The VAR statement (line 11) specifies on which variables to run PROC MEANS. PROC MEANS not only gives you means, it gives you the number of observations, the standard deviation, the minimum score found, the maximum score found, the standard error of the mean, the sum of the scores, the variance, and the coefficient of variation. (The specific statistics you get will depend on which version of SAS software you are using.) If you don't want all this information, you can specify just which pieces you want in the PROC MEANS statement. For example:

```
10      PROC MEANS N MEAN STD MAXDEC=1;
11          VAR EXAM1 EXAM2;
```

will get you just the number of cases (N), mean (MEAN), and standard deviation (STD) for the variables EXAM1 and EXAM2. In addition, the statistics will be rounded to one decimal place (because of the MAXDEC=1 option). Chapter 2 will describe most of the commonly requested options used with PROC MEANS.

The program as it is currently written will provide some useful information, but with a little more work, we can put some bells and whistles on it. The bells and whistles version below adds the following features: It computes a final grade which we will let be the average of the two exam scores; it lists the students in student number order, showing their exam scores, their final

grade and homework grade;  it computes the class average
for  the exams and final grade and a frequency count for
sex and homework grade.

```
1       DATA EXAMPLE;
2       INPUT SUBJECT SEX $ EXAM1 EXAM2 HWGRADE $;
3       FINAL = (EXAM1 + EXAM2)/2.;
4       CARDS;
5       10 M 80 84 A
6        7 M 85 89 A
7        4 F 90 86 B
8       20 M 82 85 B
9       25 F 94 94 A
10      14 F 88 84 C
11      PROC SORT;
12          BY SUBJECT;
13      PROC PRINT;
14          TITLE 'ROSTER IN STUDENT NUMBER ORDER';
15          ID SUBJECT;
16          VAR EXAM1 EXAM2 FINAL HWGRADE;
17      PROC MEANS N MEAN STD MAXDEC=1;
18          TITLE 'DESCRIPTIVE STATISTICS';
19          VAR EXAM1 EXAM2 FINAL;
```

NOTE: Once again, the line numbers to the left are
just for reference and  are not part of the program.

**Lines  1-4** constitute  our  DATA  step  or  DATA
paragraph.  **Line 1** is an instruction for the program  to
create  a  **data  set**  whose **data set name** is  "EXAMPLE."
(Remember that dataset names follow the same conventions
as variable names.)  **Line 2** tells the program  that the
first  variable in each line of data represents  SUBJECT
values,  the next variable is SEX, the third EXAM1,  and
so forth. This form of input statement is different from
that used before and requires  that  the  data values be
separated from one another by  one or more spaces,  with
data  for  each subject on a separate line. If your data
conform to this  "space-between-each-variable"  format,
then  you  don't have to specify column numbers for each
variable listed in the INPUT statement.  You may want to
anyway,  but it isn't  necessary.  (You still  have  to
follow  alphanumeric variable names with a dollar sign.)
If  you  are  going to  use  the  form  without  columns
specified,  then every variable on your data  cards must
be listed. If you specify column locations, you can pick

and  choose which variables  you desire.  **Line  3**  is  a
statement assigning the average of EXAM1  and EXAM2 to a
variable called FINAL.   The variable name "FINAL"  must
conform  to  the same naming conventions  as  the  other
variable  names  in  the INPUT  statement.  Also,  the
variable FINAL, although calculated  rather than read as
data,  is  equivalent  to the other variables  for  the
duration of the program. The "CARDS" statement in **line 4**
indicates  that  the **data step** is complete and that  the
following lines contain data.

Notice  that  each SAS statement ends with a  semicolon. As mentioned before, the semicolon is the logical
end of a SAS statement.  We could have written lines 1-4
like this:

```
DATA EXAMPLE; INPUT SUBJECT SEX $
EXAM1 EXAM2 HWGRADE $; FINAL =
(EXAM1 + EXAM2)/2.; CARDS;
```

and  the  program  would  still  run  correctly.   This
convention  is  convenient since we can write  long  SAS
statements  on several lines and simply put a  semicolon
at  the end of the statement.  **Remember to  watch  those
semicolons!**  Notice also that the data lines, since they
are  not SAS statements,  do **not** have to end with  semicolons.

Lines 5 through 10  contain our data. Remember that
if  you have data that have been placed  in  preassigned
columns with no spaces between the data values,  we must
use  the  form of the INPUT shown earlier,  with  column
specifications  after each variable name.  This form  of
data will be discussed further in Chapter 3.

Let's  spend a moment to examine what happens  when
we  execute  a SAS program.  This discussion  is  a  bit
technical  and can be skipped,  but an understanding  of
how SAS software  works will help you when you are doing
more advanced programming.  When  the  DATA statement is
executed,  SAS  software allocates a portion of a  disk
(usually  a  disk  designated  for  temporary  data--a
"scratch"  disk in computer jargon)  and names the data

set "EXAMPLE," our choice for a data set name. (More precisely, the name is "WORK.EXAMPLE" but don't worry about that for now.) The INPUT statement starts out by supplying a missing value for each of our variables. It then reads the first line of data and substitutes the actual data values for the missing values. These data values are not yet written to our SAS data set EXAMPLE but to a place called the **SAS data vector**. This is just a "holding" area where data values are stored before they get transferred to the SAS data set. The computation of the final grade comes next (line 3) and the result of this computation is added to the data vector. The CARDS line triggers the end of the data step, at which time the values in the data vector are transferred to the SAS data set. The program then returns control back to the INPUT statement to read the next line of data, compute a final grade, and write the next observation to the SAS data set. This reading, processing, and writing cycle continues until no more observations are left.

Immediately following the data is a series of PROCs. They perform various functions and computations on SAS data sets. Since we want a list of subjects and scores in subject order, we first include a SORT PROCEDURE (lines 11 and 12). Line 11 indicates that we plan to sort our data set; line 12 indicates that the sorting will be by SUBJECT number. Sorting can be multilevel if desired. For example, if we want separate lists of male and female students in subject number order, we will write:

```
PROC SORT;
    BY SEX SUBJECT;
```

This multilevel sort indicates that we should first sort by SEX (F's followed by M's--character variables are sorted alphabetically), then in SUBJECT order **within** SEX.

Lines 13 through 16 request a listing of our data (which is now in SUBJECT order). We have followed our PROC PRINT statement with three statements that supply information to the procedure. These are the **TITLE**, ID, and **VAR** statements. As with most SAS procedures, the supplementary statements following a PROC can be placed in any order. Thus

```
PROC PRINT;
   ID SUBJECT;
   TITLE 'ROSTER IN STUDENT NUMBER ORDER';
   VAR EXAM1 EXAM2 FINAL HWGRADE;
```

**is equivalent to**

```
PROC PRINT;
   TITLE 'ROSTER IN STUDENT NUMBER ORDER';
   ID SUBJECT;
   VAR EXAM1 EXAM2 FINAL HWGRADE;
```

SAS programs understand the keywords TITLE, ID, and VAR and interprets what follows in the proper context. Notice that each statement ends with its own semicolon. The words following TITLE are placed in single quotes and will be printed across the top of each of the SAS output pages. (Note: The single quotes surrounding the TITLE text are necessary for versions 5 and 6 of SAS software.) The ID variable, SUBJECT in this case, will cause the program to print the variable SUBJECT in the first column of the report, omitting the column labeled OBS (observation number) which the program will print when an ID variable is absent. The variables following the keyword VAR indicate which variables, besides the ID variable, we want in our report. The order of these variables in the list also controls the order in which they appear in the report.

Output from the complete program is shown below:

```
ROSTER IN STUDENT NUMBER ORDER   15:03 FRIDAY, JULY 25, 1986    1

SUBJECT EXAM1 EXAM2 FINAL HWGRADE

   4     90    86    88.0    B
   7     85    89    87.0    A
  10     80    84    82.0    A
  14     88    84    86.0    C
  20     82    85    83.5    B
  25     94    94    94.0    A

DESCRIPTIVE STATISTICS               15:03 FRIDAY, JULY 25, 1986    2

VARIABLE              N          MEAN         STANDARD
                                              DEVIATION

EXAM1                 6          86.5            5.2
EXAM2                 6          87.0            3.9
FINAL                 6          86.8            4.2
```

Before we leave this chapter, this is a good time to introduce you to one of the most important SAS statements--the **comment** statement. Yes, we're not kidding. A properly commented program is the sign that a true professional is at work. A **comment** inserted in a program is one or more lines of text that are **ignored** by the program--they are there only to help the programmer or researcher when he or she reads the program at a later date. To insert a comment into a SAS program, begin the comment with an asterisk (*) and end it with a semicolon. Thus,

```
*PROGRAM TO COMPUTE RELIABILITY COEFFICIENTS
RON CODY
SEPTEMBER, 1986
PROGRAM ON THE XYZ COMPUTER STORED IN ACCOUNT
3452323 WITH THE NAME FRED.
CLIENTS PHONE NUMBER IS 123-4567;
```

is a comment statement. Notice how convenient it is to include. Just enter the * and type as many lines as necessary, ending with the semicolon. You may also choose to comment individual lines in one of the following ways:

```
QUES = 6 - QUES   *TRANSFORM QUES VAR;
X = LOG(X)   *LOG TRANSFORM OF X;

        or

*TRANSFORM THE QUES VARIABLE;
QUES = 6 - QUES
*TAKE THE LOG OF X;
X = LOG(X);

        or

*
*TRANSFORM THE QUES VARIABLE
*;
QUES = 6 - QUES;
*
*TAKE THE LOG OF X
*;
X = LOG(X);
```

The last method uses more than one asterisk to set off the comment for visual effect. Note however, that each group of three lines is a **single** comment since it begins with an asterisk and ends with a semicolon.

Let us show you one final, very useful, trick using a comment statement before we conclude this chapter. Suppose you have written a program and run several procedures. Now, you come back to the program and want to run additional procedures. You could edit the program, remove the old procedures and add the new ones. Or, you could **"comment them out"** by preceding each of the program statements with an asterisk, thereby making them **comment statements**! As an example, our commented program could look like this:

```
DATA MYPROG;
INPUT X Y Z;
CARDS;
1 2 3
3 4 5
*PROC PRINT;
*    TITLE 'MY TITLE';
*    VAR X Y Z;
PROC CORR;
   VAR X Y Z;
```

The print procedure will not be executed since it will be treated as a comment; the correlation request will be run.

With most editors, adding or deleting an asterisk at the beginning of each line is very easy to do. Because of this trick (and for other reasons), we recommend that only one SAS statement be placed on a single line. If you had

```
PROC MEANS;  VAR X Y Z;
```

on one line, placing an asterisk at the beginning of the line would cause a syntax error. Please remember to comment your programs.

# Chapter 2    Describing Data

## A. DESCRIBING DATA

Even in the most complex statistical analysis, it is important to be able to describe the data in a straightforward, easy-to-comprehend fashion. This is typically accomplished in one of several ways. The first way is through descriptive summary statistics. Probably the most popular way to describe a sample of scores is by reporting: (1) the number of people in the sample (called the sample size and referred to by "n" in statistics books and SAS printouts), (2) the mean (arithmetic average) of the scores and (3) the standard deviation of the scores. The standard deviation is a measure of how widely spread the scores are. Roughly speaking, when the scores form a "bell-shaped" (normal) distribution, we expect to find about 68% of the scores to fall within 1 standard deviation of the mean (plus or minus) and to find 95% of the scores within 2 standard deviations. The steps for calculating the standard deviation are presented in the box below (for those who are interested):

---

(1) Subtract each score from the mean. The resultant numbers are called "deviations" (from the mean).

(2) Square each deviation (we now have a group of "squared deviations").

(3) Add up the squared deviations (yielding the "sum of squared deviations" or more briefly, the "sum of squares").

(4) Divide the sum of squares by the number of degrees of freedom (the number of observations minus 1). The result is called the "variance" of the scores.

(5) Take the square root of the variance and you have the **"standard deviation."**

---

Let's create a SAS data set to introduce some concepts related to descriptive statistics. Suppose we conducted a survey (albeit a very small one) where we recorded the sex, height, and weight of 7 subjects. We collected the following data:

| SEX | HEIGHT | WEIGHT |
|-----|--------|--------|
| M | 68 | 155 |
| F | 61 | 99 |
| F | 63 | 115 |
| M | 70 | 205 |
| M | 69 | 170 |
| F | 65 | 125 |
| M | 72 | 220 |

There may be several questions we want to ask about these data. Perhaps we want to count how many males and females are in our sample. We might also want means and standard deviations for the variables HEIGHT and WEIGHT. These are fairly simple tasks when only seven people are involved. However, we are rarely interested in such small samples. Once we begin to talk about as many as 20-30 people, statistical analysis by hand becomes quite tedious.

In this example, we will show the entire SAS program along with typical IBM job control language (JCL) that would be used in a typical batch environment. Here is the program:

```
1     //TEST1 JOB ACCT,NAME
2     //    EXEC  SAS                    job   control
3     //SAS.SYSIN DD *                   language
---------------------------------------------------------
4     DATA;
5     INPUT SEX $ HEIGHT WEIGHT;         SAS program
6     CARDS;                             statements
---------------------------------------------------------
7     M 68 155
8     F 61  99
9     F 63 115
10    M 70 205                           data
11    M 69 170
12    F 65 125
13    M 72 220
---------------------------------------------------------
14    PROC MEANS;                        SAS statement
---------------------------------------------------------
15    /*                                 job control
                                         language
```

Lines 1 through 3 are called job control statements. As mentioned earlier, they will vary from computer to computer. On some computer systems such as CMS or TSO, they will be replaced by terminal commands such as "filedef" etc. On microcomputers, SAS statements and Display Manager commands will take the place of these statements. In a batch environment on a mainframe computer, line 1 will, in general, contain an account number and your name. This statement is referred to as a JOB statement because it contains the word "JOB." Line 2 instructs the computer that we want to run a SAS program. It is called an EXEC statement. Finally, line 3 instructs the computer that the cards (or lines) that follow are SAS program statements and data. Since these three cards will remain the same for any batch program of this type, further examples will omit them.

If any part of this program is not clear, you should review the sample program in Chapter 1.

Let's look at the results of running this program (either by reading our card deck in a card reader, entering a command on a terminal to process our program, or submitting the program from the Display Manager on our microcomputer). Note that our output may differ slightly from the one shown below, depending on which version of SAS software you are running. As you will see in a moment, we can request specific statistics with PROC MEANS.

| VARIABLE | N | MEAN | STANDARD DEVIATION | MINIMUM VALUE | MAXIMUM VALUE |
|---|---|---|---|---|---|
| HEIGHT | 7 | 66.8571429 | 3.97611919 | 61.00000000 | 72.0000000 |
| WEIGHT | 7 | 155.5714286 | 45.79613209 | 99.00000000 | 220.0000000 |

| VARIABLE | STD ERROR OF MEAN | SUM | VARIANCE | C.V. |
|---|---|---|---|---|
| HEIGHT | 1.50283179 | 468.00000000 | 15.80952381 | 5.947 |
| WEIGHT | 17.30931093 | 1089.00000000 | 2097.28571429 | 29.437 |

For the variable of height in our sample, we see that there were seven people (n=7); that their mean height was 66.86 (rounded off); that the standard deviation was 3.98; and that the shortest person was 61 inches tall and the tallest was 72 inches (from the "minimum value" and "maximum value" columns).

In the bottom half of the table we find the standard error of the mean, the sum, the variance, and the coefficient of variation (C.V.). The sum is simply the sum of all the heights; the variance is another measure of spread, equal to the standard deviation squared. The standard error of the mean is used to put a "confidence interval" around the mean. This is useful when our scores represent a sample of scores from some population. For example, if our seven people were a random sample of high school juniors in New Jersey, we could use the sample mean (66.86) as an estimate of the average height of all New Jersey high school juniors. The standard error of the mean tells us how far off this estimate might be. If our population is roughly normally distributed, the sample estimate of the mean (based on a random sample) will fall within one standard error of the actual or "true" mean (1.50) 68% of the time and within 2 standard errors of the mean (3.00) 95% of the time. Therefore, if these scores were a random sample as described, we would be fairly confident that the population mean for New Jersey high school juniors would be between 65.36 and 68.36 (minus and plus 1 standard error) and quite confident that it would fall between 63.86 and 69.86 (minus and plus 2 standard errors). Technically speaking, this is only true when the sample is larger than 30. For smaller samples, one must use a "t" distribution table to estimate confidence intervals (see Winer).

Standard error can be calculated from the standard deviation as shown in the box below:

> The standard error is calculated by dividing the
> standard deviation by the square root of the sample
> size (n). Here: $3.976 / \sqrt{7} = 1.503$

The number under the column labeled "C.V." is the coefficient of variation. This is the standard deviation expressed as a percent of the mean. For instance, given the mean height of 66.857 and a standard deviation of 3.976, we would calculate the C.V. as 100(3.976/66.857) =5.95. The coefficient of variation is sometimes used since the magnitude of the standard deviation is only meaningful when compared to the mean. However, when the mean and standard deviation are almost always reported together, the coefficient of variation is often super-fluous.

You can specify which statistics you want to compute by specifying **options** for PROC MEANS. Most SAS procedures have **options** which are placed **between the procedure name and the semicolon.** Many of these options are listed in this text; a complete list of options for all SAS procedures can be found in the SAS manuals. As mentioned in Chapter 1, the option MAXDEC=n will control the number of decimal places for the printed statistics, N will print the number of nonmissing observations, and MEAN will produce the MEAN. So, if you want only the N and MEAN for the variables HEIGHT and WEIGHT, and you want 3 places to the right of the decimal, we would write

```
PROC MEANS N MEAN MAXDEC=3;
   VAR HEIGHT WEIGHT;
```

The **order** of the options does not matter. A list of the commonly requested options for PROC MEANS is shown next:

| OPTION | DESCRIPTION |
|--------|-------------|
| N | Number of observations on which the statistic was computed |
| NMISS | Number of missing observations |
| MEAN | Arithmetic mean |
| STD | Standard deviation |
| STDERR | Standard error |
| MIN | Minimum |
| MAX | Maximum |
| SUM | Sum |
| VAR | Variance |
| CV | Coefficient of variation |
| SKEWNESS | Skewness |
| KURTOSIS | Kurtosis |
| T | Student's t tests whether the population mean is zero. |
| PRT | The probability of obtaining a larger absolute value of t. |
| MAXDEC=n | Where n specifies the number of decimal places for printed statistics. |

PROC MEANS in the SAS/PC version produces, by default, the N, mean, standard deviation, minimum, maximum, and range. Suppose you want standard error added to the list. If you request **any** statistic (MAXDEC= is not a statistic), PROC MEANS will print **only** that statistic. Therefore, if you decide to override the system defaults and request an additional statistic, you will have to specify them **all**. As an example, with SAS/PC, to **add** standard error to the list we would write

    PROC MEANS N MEAN STD MIN MAX RANGE STDERR;        ▮


## B. FREQUENCY DISTRIBUTIONS

Let's look at how to get SAS software to count how many males and females there are in our sample. The following SAS program will do this:

```
DATA;
INPUT SEX $ HEIGHT WEIGHT;
CARDS;
M 68 155
F 61 99
F 63 115
M 70 205
M 69 170
F 65 125
M 72 220
PROC FREQ;
   TABLES SEX;
```

This time, instead of PROC MEANS, we are using a procedure (PROC) called FREQ. PROC FREQ is followed by a request for a table of frequencies for the variable SEX. In general, the word TABLES is followed by a list of variables for which we want to count occurrences of particular values (e.g., how many males and how many females for the variable SEX). Notice that the last line of the SAS program is indented several spaces from the other lines. The starting column of any SAS statement does not affect the program in any way. The last line was indented only to make it clear to the programmer that the TABLES request is part of PROC FREQ.

The table below is the output from PROC FREQ. The column labeled "FREQUENCY" lists the number of people who are males or females; the column labeled "PERCENT" is the same information expressed as a percent of the total number of people. The "CUM FREQ" and "CUM PERCENT" columns give us the cumulative counts (the number and percentage respectively) for each category of sex.

| SEX | FREQUENCY | CUM FREQ | PERCENT | CUM PERCENT |
|-----|-----------|----------|---------|-------------|
| F   | 3         | 3        | 42.857  | 42.857      |
| M   | 4         | 7        | 57.143  | 100.000     |

## C. BAR GRAPHS

We  have seen the statistics which are produced  by running PROC MEANS and PROC FREQ. It is an excellent way to get a summarization of our data.  But then, a picture is worth a thousand words (p=1000w)  so let's move on to presenting pictures of our data. SAS software can gener- ate  a frequency  bar chart showing the same information as PROC FREQ using PROC CHART.

The statements

```
PROC CHART;
   VBAR SEX;
```

were  used  to generate the frequency bar chart below:

153, 139

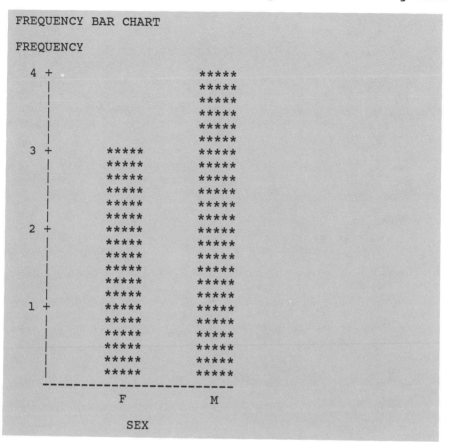

```
FREQUENCY BAR CHART

FREQUENCY

 4 +                       *****
   |                       *****
   |                       *****
   |                       *****
   |                       *****
   |                       *****
 3 +           *****       *****
   |           *****       *****
   |           *****       *****
   |           *****       *****
   |           *****       *****
   |           *****       *****
 2 +           *****       *****
   |           *****       *****
   |           *****       *****
   |           *****       *****
   |           *****       *****
   |           *****       *****
 1 +           *****       *****
   |           *****       *****
   |           *****       *****
   |           *****       *****
   |           *****       *****
   |           *****       *****
   ---------------------------------
               F           M

              SEX
```

The term HBAR in place of VBAR will generate a chart with horizontal bars instead of the vertical bars obtained from VBAR. When HBAR is used, frequency counts and percents are also presented alongside each bar (see below).

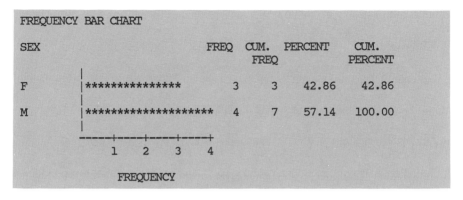

```
FREQUENCY BAR CHART

SEX                              FREQ  CUM.   PERCENT   CUM.
                                       FREQ             PERCENT

      |
F     |***************            3     3      42.86    42.86
      |
M     |********************       4     7      57.14   100.00
      |
      -----+----+----+----+
           1    2    3    4

           FREQUENCY
```

Now, what about the distribution of heights or weights? If we use PROC FREQ to calculate frequencies of heights, it will compute the number of subjects for every value of height (how many people are 60 inches tall, how many are 61 inches tall, etc.). If we use PROC CHART instead, it will automatically place the subjects into height groups (unless we specified options to control how we wanted the data displayed). Since our sample is so small, a frequency distribution of heights or weights would not be very informative. So, to demonstrate how a frequency distribution of a continuous variable like height would be displayed, a larger data set of 1000 subjects was used. The SAS statements

```
    PROC CHART;
        VBAR HEIGHT / LEVELS=20;
```

were used to generate the next chart. The option LEVELS=20 is an instruction to group the heights so that there will be 20 equally spaced intervals for the variable HEIGHT. The VBAR and HBAR statements of PROC CHART have a variety of options. The general form of the VBAR and HBAR statements is

```
VBAR variable(s) / list of options ;
```

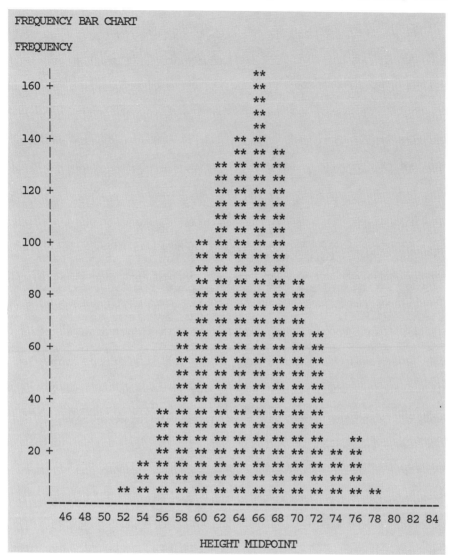

FREQUENCY BAR CHART

FREQUENCY

Notice  the shape of the distribution.  The mean is 65  inches with a standard deviation of 5 inches.  If we picture  a line drawn connecting the tops of each of the bars,  the resulting curve would approximate  a  normal curve.

Remembering our earlier discussion, approximately 68% of the people in our population would have a height of 65 inches plus or minus 5 inches (one standard deviation). This 68% also represents the proportion of the area under the entire curve from 60 inches to 70 inches.

## D. MORE DESCRIPTIVE STATISTICS AND FREQUENCY PLOTS

One procedure that can combine **descriptive statistics** with **frequency distributions** is **PROC UNIVARIATE**. This is an extremely useful procedure that can compute, among other things

1. The number of observations (nonmissing)
2. Mean
3. Standard deviation
4. Variance
5. Skewness
6. Kurtosis
7. Uncorrected and corrected sum of squares
8. Coefficient of variation
9. Standard error of the mean
10. A t-test comparing the variables' value against zero
11. Maximum (largest value)
12. Minimum (smallest value)
13. Range
14. Median, upper, and lower quartile ranges
15. Interquartile range
16. Mode
17. 1st, 5th, 10th, 90th, 95th, and 99th percentiles
18. The five highest and five lowest values (useful for data checking)
19. W or D statistic to test whether data are normally distributed
20. Stem and Leaf plot
21. Boxplot
22. Normal probability plot, comparing your cumulative frequency distribution against a normal distribution

To run PROC UNIVARIATE for our variables HEIGHT and WEIGHT, we would write

```
PROC UNIVARIATE;
    VAR HEIGHT WEIGHT;
```

To request additional options such as Stem and Leaf graphs and the test of normality we would add

```
PROC UNIVARIATE NORMAL PLOT;
    VAR HEIGHT WEIGHT;
```

A portion of the output from the above request is shown next:

```
UNIVARIATE

Variable=HEIGHT

                    Moments

N                 7   Sum Wgts            7
Mean       66.85714   Sum               468
Std Dev    3.976119   Variance    15.80952
Skewness   -.320437   Kurtosis    -1.23625
USS           31384   CSS         94.85714
CV         5.947187   Std Mean    1.502832
T:Mean=0   44.48744   Prob>|T|      0.0001
Sgn Rank         14   Prob>|S|      0.0225
Num ^= 0          7
W:Normal   .9571652   Prob<W         0.805

           Quantiles(Def=5)                      Extremes

100% Max          72    99%        72    Lowest   Highest
 75% Q3           70    95%        72        61        65
 50% Med          68    90%        72        63        68
 25% Q1           63    10%        61        65        69
  0% Min          61     5%        61        68        70
                         1%        61        69        72
Range             11
Q3-Q1              7
Mode              61
```

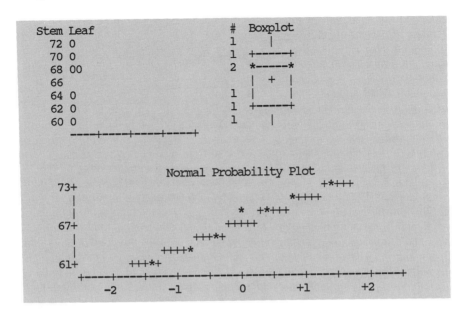

```
Stem Leaf                        #  Boxplot
  72 0                           1   |
  70 0                           1  +-----+
  68 00                          2  *-----*
  66                                |  +  |
  64 0                           1  |     |
  62 0                           1  +-----+
  60 0                           1     |
     ----+----+----+----+
```

```
                   Normal Probability Plot
  73+                                            +*+++
    |                                       *+++
    |                               *  +*+++
  67+                            +++++
    |                        +++*+
    |                    ++++*
  61+             +++*+
     +----+----+----+----+----+----+----+----+----+
         -2        -1        0        +1        +2
```

To summarize what we've done up to this point:

(1)  We can get summary statistics by using PROC
MEANS  and listing the variables we want to  look
at.

          PROC MEANS;

will  produce summary statistics for all  variables
listed as numeric in the data set.

          PROC MEANS;
             VAR variable(s);

will produce summary statistics only for those variables listed.

(2)  We  can get frequency tables for any numeric
variable by using

          PROC FREQ;
             TABLES variable(s);

(3) We can get histograms by using PROC CHART.

```
PROC CHART;
     VBAR variable(s);
```

will produce vertical charts and

```
PROC CHART;
     HBAR variable(s);
```

will produce horizontal charts AND frequencies.

(4) We can produce descriptive statistics and frequency plots, using PROC UNIVARIATE.

```
PROC UNIVARIATE;
     VAR variable(s);
```

## E. PLOTTING DATA

We would now like to investigate the relationship between height and weight. Our intuition tells us that these two variables are related: the taller a person is, the heavier (in general). The best way to display this relationship is to draw a graph of height versus weight. We can have our SAS program generate this graph by using PROC PLOT. The statements

```
PROC PLOT;
     PLOT WEIGHT*HEIGHT;
```

will generate the graph that follows:  (Note:  This is a plot using the **original** data set of seven people.)

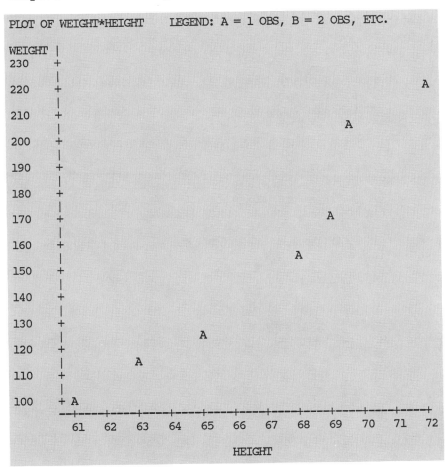

```
PLOT OF WEIGHT*HEIGHT      LEGEND: A = 1 OBS, B = 2 OBS, ETC.

WEIGHT |
230    +
       |
220    +                                                          A
       |
210    +
       |                                            A
200    +
       |
190    +
       |
180    +
       |
170    +                                      A
       |
160    +                                 A
       |
150    +
       |
140    +
       |
130    +
       |                    A
120    +
       |         A
110    +
       |
100    + A
       --+----+----+----+----+----+----+----+----+----+----+----+
         61   62   63   64   65   66   67   68   69   70   71   72
                                 HEIGHT
```

The general form of the plot command is

```
PROC PLOT;
    PLOT Y variable * X variable;
        (Vertical)     (Horizontal)
```

Notice that SAS software automatically chooses appropriate scales for the X and Y axes. Unless you specify otherwise, PROC PLOT uses letters (A,B,C, etc.) as plotting symbols. Since a computer line printer is restricted to printing characters in discrete locations across or down the page, two data values that are very close to each other would need to be printed in the same location. If **two** data points do occur at one print position, the program prints the letter "B"; for **three** data points, the letter "C," and so forth.

So far we have looked at descriptive statistics (mean, standard deviation, standard error, etc.) for height and weight, we have counted the number of males and females, and we have seen the relationship between height and weight in a graph. Another useful way of looking at our data comes to mind. Can we calculate descriptive statistics separately for the males and females? Can we obtain a plot of height versus weight for males, and one for females? The answer is "yes," and it is quite easy to do. We must first ask the SAS program to group our data by SEX. Once this is done, we can use PROC MEANS and PROC PLOT to produce the desired statistics and graphs.

Our program will look as follows:

```
DATA;
INPUT SEX $ HEIGHT WEIGHT;
CARDS;
M 68 155
F 61 99
F 63 115
M 70 205
M 69 170
F 65 125
M 72 220
PROC SORT;
   BY SEX;
PROC MEANS;
   BY SEX;
PROC PLOT;
   BY SEX;
   PLOT WEIGHT*HEIGHT;
```

The PROC SORT statements will arrange our data so that they will be grouped by SEX. Once the data set has been sorted, we now have the option to use a **BY** statement with PROC MEANS or PROC PLOT. The result of this program will be a separate analysis for males and another for females. If we omit a BY statement with PROC MEANS or PROC PLOT, the program will ignore the fact that the data set is now grouped with a BY variable.

We  can generate one final graph that will  display
the  data  for males and females on a single graph  but,
instead of the usual plotting symbols of A,B,C, etc., we
will use F's and M's (for females and males). The statements

```
PROC PLOT;
    PLOT WEIGHT*HEIGHT=SEX;
```

will  accomplish this.  The data set does not have to be
sorted to use this form of PROC PLOT.  The "=SEX"  after
our plot request specifies that the first letter of each
of the SEX values will be used as plotting symbols.   In
essence,  this allows us to look at 3 variables (height,
weight, and sex) at the same time. The result of running
this last procedure is shown below:

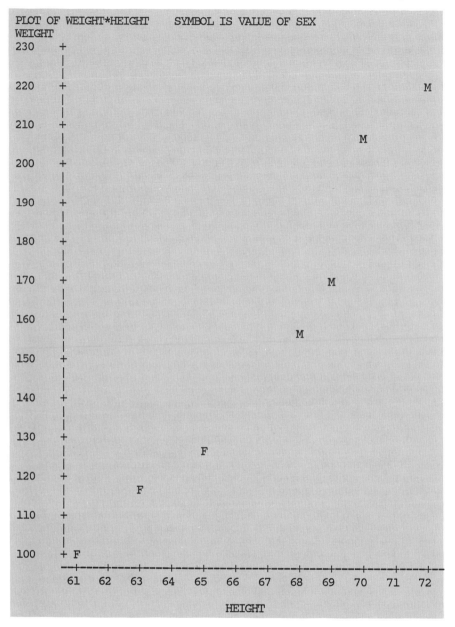

```
PLOT OF WEIGHT*HEIGHT      SYMBOL IS VALUE OF SEX
WEIGHT
230    +
       |
       |
220    +                                                          M
       |
       |
210    +
       |                                              M
       |
200    +
       |
       |
190    +
       |
       |
180    +
       |
       |
170    +                                     M
       |
       |
160    +
       |                                 M
       |
150    +
       |
       |
140    +
       |
       |
130    +
       |                      F
       |
120    +
       |            F
       |
110    +
       |
       |
100    + F
     --+----+----+----+----+----+----+----+----+----+----+----+--
       61   62   63   64   65   66   67   68   69   70   71   72
                            HEIGHT
```

Since we are not using the standard plotting symbols (A, B, C, etc.), multiple observations at a single print location will not be shown on the graph (except in the case of one male and one female, in which case the M and F will overprint). The program will print a message indicating the number of "hidden" observations at the bottom of the graph in this case.

# Chapter 3    Questionnaire Design and Analysis

## A. SURVEY DATA

A common way of collecting certain types of data is with a questionnaire. Although these can be designed in many ways, the following example contains features that make it especially useful when the collected data are to be entered into a computer.

--------------------------------------------------------------------

| | For office use only |
|---|---|
| **Sample Questionnaire** | ID  |__|__|__| |

1. Age in years _____                          |__|__|

2. Sex  __1 Male
        __2=Female                                        |__|

3. Race __1=White
        __2=Black
        __3=Hispanic
        __4=Other                                         |__|

4. Marital status
        __1=Single
        __2=Married
        __3=Widowed
        __4=Divorced                                      |__|

5. Education level
        __1=High school or less
        __2=Two year college
        __3=Four year college (B.A. or B.S.)
        __4=Post graduate degrees                         |__|

For each of the following statements, please place the NUMBER that corresponds with your feelings to the left of the question number. Use the following codes:

1=Strongly disagree 2=Disagree 3=Neutral 4=Agree
5=Strongly agree

____6. The president of the U.S. has been doing a good job.   |__|

____7. The arms budget should be increased.                  |__|

____8. There should be more federal aid to big cities.       |__|

Notice that every response in this questionnaire is placed in a box by a coding clerk and that the boxes are all on the right side of the page. This will facilitate the job of transferring the data from the survey instrument to our computer. One should be careful, however, not to ignore the person who is filling out the questionnaire. If the questionnaire confuses the respondent, it will not matter how easy the data are to enter. With this in mind, many experienced questionnaire designers would place the choices for questions 6 through 8 below each of these items and have the respondent check his choice. The typical way of coding data from a questionnaire of this type would be to have the data either keypunched on cards, entered into a computer using a terminal, or entered into a microcomputer using a database management system that can output an ASCII text file. In either case, we would probably set aside certain card columns (or positions on a line) for each variable. The data can either be entered directly from the questionnaire or transcribed to a coding form. Usual coding forms are ruled into 80 columns.

We might decide to enter our questionnaire data on cards as follows:

| Card Column | Description | Variable Name |
|---|---|---|
| 1-3 | Subject ID | ID |
| 4-5 | Age in years | AGE |
| 6 | Sex | SEX |
| 7 | Race | RACE |
| 8 | Marital status | MARITAL |
| 9 | Education level | EDUC |
| 10 | President doing good job | PRES |
| 11 | Arms budget increased | ARMS |
| 12 | Federal aid to cities | CITIES |

Typical cards of data would look like this:

```
000091113232
002452222422
```

Notice that we have not left any spaces between the values for each variable. Therefore, we **must** specify the column location for each variable. Our INPUT statement for this questionnaire would be written:

```
INPUT ID 1-3 AGE 4-5 SEX 6 RACE 7 MARITAL 8 EDUC 9
     PRES 10 ARMS 11 CITIES 12;
```

Each variable name is followed by its column designation. A common occurrence with questionnaires is that some people will not answer all the questions. With a **list** INPUT statement (one in which we list only the variable names and not the column designations) we use a **period** to represent a **missing value**; with our **column** INPUT statement we leave the column(s) **blank**. We can do this since it is the **columns**, not the order of the data, that determine which variable is being read.

A complete SAS program to (1) calculate the mean age of the respondents and (2) compute frequencies for all the other variables is shown below:

```
DATA;
INPUT ID 1-3 AGE 4-5 SEX 6 RACE 7 MARITAL 8 EDUC 9
      PRES 10 ARMS 11 CITIES 12;
CARDS;
000091113232
002452222422
003351324442
004271111121
005682132333
006651243425
    etc.
PROC MEANS MAXDEC=2;
  VAR AGE;
PROC FREQ;
    TABLES SEX RACE MARITAL EDUC PRES ARMS CITIES;
```

The PROC MEANS statement has the option MAXDEC=2 (maximum number of decimals=2) added. This will cause all the PROC MEANS statistics to be rounded to 2 places to the right of the decimal point. The VAR AGE part of PROC MEANS gives instructions that we want to calculate statistics for the variable AGE. Remember, without the "VAR" statement, PROC MEANS would calculate means for all numeric variables in the data set. The general form of the PROC MEANS statement is

```
PROC MEANS  options ;
  VAR  variable(s) ;
```

where the word "options" will be replaced by a list of valid options taken from the SAS manual and "variable(s)" is a single variable name or a list of variable names, separated by spaces.

We could also save some time by writing our TABLES request like this:

```
TABLES SEX -- CITIES;
```

This convention "variable name-- variable name" means to include all the variables from SEX to CITIES in the order they exist in the data set. In this case, the order is the same as the order on the INPUT statement. (See Section E of Reference Chapter 2 for more details.) Notice that we have not requested any statistics for the variable ID. The ID number only serves as an identifier if we want to go back to the original questionnaire to check data values.

A sample of the output from PROC FREQ is shown below:

| SEX | FREQUENCY | CUM FREQ | PERCENT | CUM PERCENT |
|---|---|---|---|---|
| 1 | 4 | 4 | 66.667 | 66.667 |
| 2 | 2 | 6 | 33.333 | 100.000 |

| RACE | FREQUENCY | CUM FREQ | PERCENT | CUM PERCENT |
|---|---|---|---|---|
| 1 | 3 | 3 | 50.000 | 50.000 |
| 2 | 2 | 5 | 33.333 | 83.333 |
| 3 | 1 | 6 | 16.667 | 100.000 |

| MARITAL | FREQUENCY | CUM FREQ | PERCENT | CUM PERCENT |
|---|---|---|---|---|
| 1 | 2 | 2 | 33.333 | 33.333 |
| 2 | 2 | 4 | 33.333 | 66.667 |
| 3 | 1 | 5 | 16.667 | 83.333 |
| 4 | 1 | 6 | 16.667 | 100.000 |

We could improve on this output considerably. First, we have to refer back to our coding scheme to see the definition of each of the variable names. Some variable names like SEX and RACE need no explanation; others like PRES and CITIES do. We can associate a **VARIABLE LABEL** with each **VARIABLE NAME** by using a **LABEL** statement. These labels will be printed along with the variable name in certain procedures such as PROC FREQ and PROC MEANS. The general form of a LABEL statement is

    LABEL variable name=description
                   .
                   .
                   .
          variable name=description;

The "description" can contain up to 40 characters (blanks count as characters) and must be enclosed in single quotes (new with version 6 of SAS software). The LABEL statement will usually be placed before the CARDS statement. Our program, rewritten to include variable labels, follows:

```
DATA;
INPUT ID 1-3 AGE 4-5 SEX 6 RACE 7 MARITAL 8 EDUC 9
     PRES 10 ARMS 11 CITIES 12;
LABEL  MARITAL='MARITAL STATUS'
       EDUC='EDUCATION LEVEL'
       PRES='PRESIDENT DOING A GOOD JOB'
       ARMS='ARMS BUDGET INCREASE'
       CITIES='FEDERAL AID TO CITIES';
CARDS;
000091113232
002452222422
003351324442
004271111121
005682132333
006651243425
PROC MEANS MAXDEC=2;
   VAR AGE;
PROC FREQ;
     TABLES SEX RACE MARITAL EDUC PRES ARMS CITIES;
```

Notice that we did not supply a variable label  for
**all**  our variables.  The ones you choose are up to  you.
Now when we run our program,  the labels will be printed
along with the variable names in our PROC FREQ output. A
sample of the output from this program is shown below:

| SEX | FREQUENCY | CUM FREQ | PERCENT | CUM PERCENT |
|---|---|---|---|---|
| 1 | 4 | 4 | 66.667 | 66.667 |
| 2 | 2 | 6 | 33.333 | 100.000 |

| RACE | FREQUENCY | CUM FREQ | PERCENT | CUM PERCENT |
|---|---|---|---|---|
| 1 | 3 | 3 | 50.000 | 50.000 |
| 2 | 2 | 5 | 33.333 | 83.333 |
| 3 | 1 | 6 | 16.667 | 100.000 |

MARITAL STATUS

| MARITAL | FREQUENCY | CUM FREQ | PERCENT | CUM PERCENT |
|---|---|---|---|---|
| 1 | 2 | 2 | 33.333 | 33.333 |
| 2 | 2 | 4 | 33.333 | 66.667 |
| 3 | 1 | 5 | 16.667 | 83.333 |
| 4 | 1 | 6 | 16.667 | 100.000 |

```
EDUCATION LEVEL
EDUC    FREQUENCY   CUM FREQ    PERCENT   CUM PERCENT

  1          1          1       16.667      16.667
  2          2          3       33.333      50.000
  3          2          5       33.333      83.333
  4          1          6       16.667     100.000

PRESIDENT DOING A GOOD JOB
PRES    FREQUENCY   CUM FREQ    PERCENT   CUM PERCENT

  1          1          1       16.667      16.667
  2          1          2       16.667      33.333
  3          1          3       16.667      50.000
  4          3          6       50.000     100.000

ARMS BUDGET INCREASE
ARMS    FREQUENCY   CUM FREQ    PERCENT   CUM PERCENT

  2          3          3       50.000      50.000
  3          2          5       33.333      83.333
  4          1          6       16.667     100.000

FEDERAL AID TO CITIES
CITIES    FREQUENCY   CUM FREQ   PERCENT   CUM PERCENT

  1          1          1       16.667      16.667
  2          3          4       50.000      66.667
  3          1          5       16.667      83.333
  5          1          6       16.667     100.000
```

We would like to improve the readability of the output one step further. The remaining problem is that the values for our variables (1=male 2=female etc.) are printed on the output, not the names that we have assigned to these values. We would like the output to show the number of males and females, for example, not the number of 1's and 2's for the variable SEX. We can supply the VALUE LABELS in two steps.

The first step is to define our code values for each variable. For example, 1=male, 2=female will be used for our variable SEX. The codes 1=str disagree, 2=disagree, etc. will be used for three variables: PRES, ARMS, and CITIES. SAS software calls these codes FORMATS and we define the FORMATS in a procedure by that name.

The second step, shown later, will be to associate a FORMAT with one or more variable names. Below is an example of PROC FORMAT used for our questionnaire:

```
PROC FORMAT;
   VALUE SEXFMT 1='MALE' 2='FEMALE';
   VALUE RACEFMT   1='WHITE'  2='BLACK' 3='HISPANIC'
                   4='OTHER';
   VALUE MARFMT   1='SINGLE' 2='MARRIED' 3='WIDOWED'
                   4='DIVORCED';
   VALUE EDUCFMT 1='HIGH SCH OR LESS'
                 2='TWO YR. COLLEGE'
                 3='FOUR YR. COLLEGE'
                 4='GRADUATE DEGREE';
   VALUE LIKERT 1='STR DISAGREE'
                2='DISAGREE'
                3='NEUTRAL'
                4='AGREE'
                5='STR AGREE';
```

The names SEXFMT, RACEFMT, LIKERT, etc. are format names. (An aside: We chose the name LIKERT since scales such as 1=strongly disagree, 2=disagree, 3=neutral etc. are called Likert scales by psychometricians.) You may choose any name (consistent with naming conventions, with the exception that they cannot end in a number) for your formats. It is best to give names that help you remember what the format will be used for. As a matter of fact, **you can use the same name for a format and for a variable** without confusion. Thus, the format for SEX could be called "SEX." Formats for character variables must start with a $ sign. If we had coded SEX as a character variable using M's and F's for Male and Female, a format for this variable could be created with:

```
VALUE $SEX 'M'='MALE'    'F'='FEMALE';
```

Descriptions (or value labels as some people call them), can be up to 40 characters long but should be restricted to 16 characters since this is the number of characters that will appear in a cross tabulation or frequency table. They are placed after the equal sign and must be placed in single quotes.

Once we have defined a set of formats (such as SEXFMT, RACEFMT, LIKERT, etc.), we need to assign the formats to the appropriate variables. Just because we have named a format SEXFMT, for example, does not mean that this format will be used with the variable we have called SEX. We need another SAS statement that indicates which formats will be used or associated with which variables. Format statements start with the word FORMAT, followed by a single variable or a list of variables, followed by a format name to be used with the preceding variables. The list of variable and format names continues on as many lines as necessary and ends with a semicolon. SAS software knows the difference between our **variable** names and our **format** names since we place a PERIOD AFTER EACH FORMAT NAME IN OUR FORMAT STATEMENT. Format statements can be placed within a **DATA** step or as a statement in a **PROC** step. If we choose to place our format statement in the DATA step, the formatted values will be associated with the assigned variable(s) **for all** PROCS that use the data set. **If we place a format** statement in a PROC step, the formatted values will be used **only for that procedure.** In this example we will place our format statement in the DATA step. Therefore, we will define our formats with PROC FORMAT **before** we write our DATA step. We will associate the format SEXFMT with the variable SEX, RACEFMT with the variable RACE, and so forth. The variables PRES, ARMS, and CITIES are all "sharing" the LIKERT format. If you look at the completed questionnaire program below, the use of PROC FORMAT and its application in other PROCs should become clear.

```
PROC FORMAT;
    VALUE SEXFMT 1='MALE' 2='FEMALE';
    VALUE RACEFMT 1='WHITE'2='BLACK'3='HISPANIC'4='OTHER';
    VALUE MARFMT 1='SINGLE' 2='MARRIED' 3='WIDOWED'
                4='DIVORCED';
    VALUE EDUCFMT 1='HIGH SCH OR LESS'
                  2='TWO YR. COLLEGE'
                  3='FOUR YR. COLLEGE'
                  4='GRADUATE DEGREE';
    VALUE LIKERT 1='STR DISAGREE'
                 2='DISAGREE'
                 3='NEUTRAL'
                 4='AGREE'
                 5='STR AGREE';
DATA;
INPUT ID 1-3 AGE 4-5 SEX 6 RACE 7 MARITAL 8 EDUC 9
      PRES 10 ARMS 11 CITIES 12;
LABEL  MARITAL='MARITAL STATUS'
       EDUC='EDUCATION LEVEL'
       PRES='PRESIDENT DOING A GOOD JOB'
       ARMS='ARMS BUDGET INCREASE'
       CITIES='FEDERAL AID TO CITIES';
FORMAT SEX SEXFMT. RACE RACEFMT. MARITAL MARFMT.
       EDUC EDUCFMT. PRES ARMS CITIES LIKERT.;
CARDS;
000091113232
002452222422
003351324442
004271111121
005682132333
006651243425
PROC MEANS MAXDEC=2;
    VAR AGE;
PROC FREQ;
      TABLES SEX RACE MARITAL EDUC PRES ARMS CITIES;
```

Output from PROC FREQ in this program is shown below:

| SEX | FREQUENCY | CUM FREQ | PERCENT | CUM PERCENT |
|---|---|---|---|---|
| MALE | 4 | 4 | 66.667 | 66.667 |
| FEMALE | 2 | 6 | 33.333 | 100.000 |
| | | | | |
| RACE | FREQUENCY | CUM FREQ | PERCENT | CUM PERCENT |
| WHITE | 3 | 3 | 50.000 | 50.000 |
| BLACK | 2 | 5 | 33.333 | 83.333 |
| HISPANIC | 1 | 6 | 16.667 | 100.000 |

```
MARITAL STATUS
MARITAL     FREQUENCY   CUM FREQ     PERCENT    CUM PERCENT

SINGLE          2           2        33.333        33.333
MARRIED         2           4        33.333        66.667
WIDOWED         1           5        16.667        83.333
DIVORCED        1           6        16.667       100.000


EDUCATION LEVEL
EDUC               FREQUENCY  CUM FREQ    PERCENT  CUM PERCENT

HIGH SCH OR LESS       1         1        16.667      16.667
TWO YR. COLLEGE        2         3        33.333      50.000
FOUR YR. COLLEGE       2         5        33.333      83.333
GRADUATE DEGREE        1         6        16.667     100.000


PRESIDENT DOING A GOOD JOB
PRES            FREQUENCY  CUM FREQ     PERCENT   CUM PERCENT

STR DISAGREE        1         1         16.667       16.667
DISAGREE            1         2         16.667       33.333
NEUTRAL             1         3         16.667       50.000
AGREE               3         6         50.000      100.000


ARMS BUDGET INCREASE
ARMS        FREQUENCY   CUM FREQ     PERCENT    CUM PERCENT

DISAGREE        3           3         50.000        50.000
NEUTRAL         2           5         33.333        83.333
AGREE           1           6         16.667       100.000


FEDERAL AID TO CITIES
CITIES          FREQUENCY  CUM FREQ     PERCENT   CUM PERCENT

STR DISAGREE        1         1         16.667       16.667
DISAGREE            3         4         50.000       66.667
NEUTRAL             1         5         16.667       83.333
STR AGREE           1         6         16.667      100.000
```

## B. RECODING DATA

In the previous questionnaire example, we coded the
respondent's actual age in years. What if we want to
look at the relationship between age and the questions
6-8 (opinion questions)? It might be convenient to have
a variable that indicated an age group rather than the
person's actual age. We will look at two ways of accom-
plishing this.

Look at the following SAS statements:

```
PROC FORMAT;
   VALUE SEXFMT 1=MALE 2=FEMALE;
        .
        .
        .
   VALUE AGEFMT 1='0-20'  2='21-40'  3='41-60'
                4='GREATER THAN 60';
DATA;
INPUT ID 1-3 AGE 4-5 SEX 6 RACE 7 MARITAL 8 EDUC 9
      PRES 10 ARMS 11 CITIES 12;
IF 0 < AGE <=20 THEN AGEGRP=1;
IF 20 < AGE <= 40 THEN AGEGRP=2;
IF 40 < AGE <= 60 THEN AGEGRP=3;
IF AGE > 60 THEN AGEGRP=4;
LABEL  MARITAL='MARITAL STATUS'
        .
        .
        .
      AGEGRP='AGE GROUP';
   FORMAT SEX SEXFMT. RACE RACEFMT. MARITAL MARFMT.
          EDUC EDUCFMT. PRES ARMS CITIES LIKERT.
          AGEGRP AGEFMT.;
CARDS;
   (data)
PROC FREQ;
   TABLES SEX -- AGEGRP;
```

Several new features have been added to the program. The major additions are the four IF statements following the INPUT. With the DATA statement, the program begins to create a data set. When the INPUT step is reached, the program will read a line of data according to the INPUT specifications. Next, each IF statement is evaluated. If the condition is TRUE, then the variable AGEGRP will be set to 1,2,3, or 4. Finally, when the CARDS statement is reached, an observation is added to the SAS data set. The variables in the data set will include all the variables listed in the INPUT statement as well as the variable AGEGRP. The variable AGEGRP may be used in any PROC just like any of the other variables. Be sure there are no "cracks" in your recoding ranges. That is, make sure you code your IF statements so that there isn't a value of AGE that is not recoded. If that happens, the variable AGEGRP for that person will have a **missing value**. A better way to write multiple IF statements is to use an **ELSE** before all but the first IF. The four IF statements would then look like this:

```
IF 0 < AGE <=20 THEN AGEGRP=1;
   ELSE IF 20 < AGE <= 40 THEN AGEGRP=2;
   ELSE IF 40 < AGE <= 60 THEN AGEGRP=3;
   ELSE IF AGE > 60 THEN AGEGRP=4;
```

The effect of the ELSE statements is that when any IF statement is **true**, all the following ELSE statements will be skipped. The advantage is to reduce computer time (since all the IF do not have to be tested) and to avoid the following type of problem. Can you see what will happen with the statements below?

(Assume X can have values of 1,2,3,4, or 5)

```
IF X=1 THEN X=5;
IF X=2 THEN X=4;
IF X=4 THEN X=2;
IF X=5 THEN X=1;
```

What happens when X is 1? The first IF statement is true. This causes X to have a value of 5. The next two IF statements are false but the last IF statement is **true**. X is back to 1! The ELSE statements, besides reducing computer time, prevent the problem above. The correct coding is

```
IF X=1 THEN X=5;
   ELSE IF X=2 THEN X=4;
   ELSE IF X=4 THEN X=2;
   ELSE IF X=5 THEN X=1;
```

One final note: if all we wanted to do was recode X so that 1=5, 2=4, 3=3, 4=2, and 5=1 the statement

```
X = 6 - X;
```

would be the best way to recode the X values.

Notice that we added a line to the LABEL section and to PROC FORMAT to supply a variable label and a format for our new variable.

There is another way of recoding our AGE variable without creating a new variable. We will use a "trick." By defining a special format, we can have SAS software assign subjects to age categories. We can write

```
PROC FORMAT;
   VALUE AGROUP 0-20='0-20'
               21-40='21-40'
               41-60='41-60'
               60-HIGH=GREATER THAN 60;
```

Instead of single values to the left of the = sign, we are supplying a range of values. The special words HIGH and LOW are available to indicate all values under or over a specified value. (We used the range 0-20 instead of LOW-20 since the designation "LOW-20" will include missing values that are stored in the computer as large negative numbers.)

Once we have defined a format, we can then issue a TABLES request on the original variable (such as AGE). But, by supplying the format information using the format AGROUP for the variable AGE, the new categories will be printed instead of the original AGE values. Notice that we will now place the format statement in the appropriate PROC rather than in the DATA statement since we want to use the recoded values only for PROC FREQ.

Thus, the SAS statements

```
PROC FREQ;
   TABLES AGE;
   FORMAT AGE AGROUP.;
```

will produce the following output:

| AGE | FREQUENCY | CUM FREQ | PERCENT | CUM PERCENT |
|---|---|---|---|---|
| 21-40 | 3 | 3 | 50.000 | 50.000 |
| 41-60 | 1 | 4 | 16.667 | 66.667 |
| GREATER THAN 60 | 2 | 6 | 33.333 | 100.000 |

## C. TWO-WAY FREQUENCY TABLES

Besides computing frequencies on individual variables, we might have occasion to count occurrences of
one variable at each level of another variable. An example will make this clear. Suppose we took a poll of
presidential preference and also recorded the sex of the
respondent. Sample data might look like this:

```
     SEX     CANDIDATE
---------------------------
      M       DEWEY
      F       TRUMAN
      M       TRUMAN
      M       DEWEY
      F       TRUMAN
         etc.
```

We would like to know (1) how many people were for
Dewey and how many for Truman, (2) how many males and
females were in the sample, and (3) how many males and
females were for Dewey and Truman, respectively.

A previous example of PROC FREQ shows how to perform tasks (1) and (2). For (3) we would like a table
that looks like this:

```
                        SEX

                 F    |    M
             ------------------
             |        |        |
  DEWEY      |   70   |   40   |  110
             |_____|_____|
             |        |        |
  TRUMAN     |   30   |   40   |   70
             |_____|_____|

                100       80      180
```

If this were our table, it would show that females favored Dewey over Truman 70 to 30 while males were split evenly. A SAS program to solve all three tasks is given below:

```
DATA;
INPUT SEX $ CANDID $ ;
CARDS;
M DEWEY
F TRUMAN
M TRUMAN
M DEWEY
F TRUMAN
   etc.
PROC FREQ;
    TABLES SEX CANDID;
    TABLES CANDID*SEX;
```

Notice that since the variables SEX and CANDID are coded as character values (alphanumeric), we follow each variable name with a $ in the INPUT statement. Another fact that we have not mentioned so far is that the VALUES of our character variables also cannot be longer than eight letters in length unless we modify our INPUT statement to indicate this. So, for the time being, we cannot use this program for the Eisenhower/Stevenson election (without using nicknames).

The first TABLES request is the same type we have seen before; the second is a request for a two-way table.

What would a table like the one above tell us? If it were based on a random sample of voters, we might conclude that sex affected voting patterns. Before we conclude that this is true of the nation as a whole, it would be nice to see how likely it was that these results were simply due to a quirky sample. A statistic called chi-square ($x^2$) will do just this.

Consider the table again. There were 180 people in our sample, 110 for Dewey and 70 for Truman; 100 females and 80 males. If there were no sex bias, we would expect the proportion of the population who wanted Dewey (110/180) to be the same for the females and males. Therefore, since there were 100 females, we could expect (110/180) of 100 (approximately 61) females to be for Dewey. Our expectations (in statistics called **"expected values"**) for all the other cells can be calculated in the same manner. Once we have **observed** and **expected** frequencies for each cell (each combination of sex and candidate), the chi-square statistic can be computed. By adding an option for chi-square on our TABLES request, we can have our program compute chi-square and the probability of obtaining a value as large or larger by chance alone. The modified TABLES request is written:

▌    TABLES CANDID*SEX / CHISQ;

Output from the above request is shown below:

```
TABLE OF CANDID BY SEX
CANDID          SEX

FREQUENCY|
 PERCENT |
 ROW PCT |
 COL PCT |     F    |    M    |  TOTAL
---------+--------+--------+
DEWEY     |     70 |     40 |    110
          |  38.89 |  22.22 |  61.11
          |  63.64 |  36.36 |
          |  70.00 |  50.00 |
---------+--------+--------+
TRUMAN    |     30 |     40 |     70
          |  16.67 |  22.22 |  38.89
          |  42.86 |  57.14 |
          |  30.00 |  50.00 |
---------+--------+--------+
TOTAL           100       80       180
                55.56    44.44   100.00
STATISTICS FOR 2-WAY TABLES
CHI-SQUARE                     7.481   DF=  1   PROB=0.0062
PHI                            0.204
CONTINGENCY COEFFICIENT        0.200
CRAMER'S V                     0.204
LIKELIHOOD RATIO CHI-SQUARE    7.493   DF=  1   PROB=0.0062
CONTINUITY ADJ. CHI-SQUARE     6.663   DF=  1   PROB=0.0098
FISHER'S EXACT TEST (1-TAIL)                    PROB=0.0049
                    (2-TAIL)                    PROB=0.0087
```

The key to the table is found in the upper left-hand corner of the table. By FREQUENCY, we mean the number of subjects in the cell. For example, 70 females favored Dewey for president. The second number in each cell shows the PERCENT of the **total population**. The third number, labeled ROW PCT gives the percent of **each** row. **For example, of all the people for Dewey (row 1),** 70/110 x 100 or 63.64% were female. The last number, COL PCT, is the **column percent**. Of all the **females**, 70% were for Dewey and 30% were for Truman. It is customary to place the variable that we consider the independent or causal variable (SEX in our example) along the columns of the table. In the TABLES request for a two-way cross tabulation, the variable that forms the columns is placed second (e.g., CANDID*SEX). In our statistical requests, rows come first, then columns.

For our example, chi-square equals 7.48 and the probability of obtaining a chi-square this large or larger by chance alone is .006. Therefore, we can say that based on our data, there is a sex bias in presidential preference: there is a tendency for females to show greater preference for Dewey than males do.

The number of degrees of freedom (df) in a chi-square statistic is equal to the number of rows minus one multiplied by the number of columns minus one ((R-1)x(C-1)). Thus, our 2x2 chi-square has 1 df. If there is only one row in the table, then the df are simply the number of cells minus one. Whenever a chi-square table has 1 df and the **expected** value of any cell is less than 5, a "correction for continuity" called Yates' correction should be applied. SAS software prints out a corrected chi-square value and its associated probability beside the heading CONTINUITY ADJ. CHI-SQUARE. When df are greater than 1, no more than 20% of the cells should have **expected** values less than 5. The program will print a warning when this condition occurs. This does not mean that you have to throw your data out if you fall into this situation. Below, we will tell you one alternative if your df is greater than 1. If your df=1, use either the corrected chi-square or Fisher's exact test, which is also printed in the SAS output when df=1. If you are in doubt, consult your local statistician.

For larger tables (more than four cells) the usual alternative when faced with small **expected** cell values is to combine or collapse cells. If we had four categories of age: 0-20, 21-40, 41-60, and over 60, we might combine 0-20 and 21-40 as one group, and 41-60 and 60+ as another. Another example would be combining categories such as "strongly disagree" and "disagree" on an opinion questionnaire. We can use either method of recoding shown in the previous section to accomplish this.

We can use the questionnaire program in sections A and B of this chapter to see another example of a two-way table. Supose we wanted cross-tabulations of AGEGRP against the three variables PRES, ARMS, and CITIES. We could code

      TABLES (PRES ARMS CITIES)*AGEGRP;                    ∎


This will generate three tables and is a short way of writing

      TABLES PRES*AGEGRP ARMS*AGEGRP CITIES*AGEGRP;        ∎

We can also have multiple column variables in a TABLE request. Thus

      TABLES (PRES ARMS)*(AGEGRP SEX);                     ∎

would produce four tables PRES*AGEGRP, PRES*SEX, ARMS*AGEGRP, and ARMS*SEX.

When you use this method, be sure to put the list of variables in parentheses.

One of the tables generated from this program is shown below:

```
TABLE OF PRES BY AGE
PRES          PRESIDENT DOING A GOOD JOB     AGE

FREQUENCY    |
 PERCENT     |
 ROW PCT     |
 COL PCT     |21-40    |41-60    |GREATER  |
             |         |         |THAN 60  |   TOTAL
-------------+---------+---------+---------+
STR DISAGREE |      1  |      0  |      0  |       1
             |  16.67  |   0.00  |   0.00  |   16.67
             | 100.00  |   0.00  |   0.00  |
             |  33.33  |   0.00  |   0.00  |
-------------+---------+---------+---------+
DISAGREE     |      1  |      0  |      0  |       1
             |  16.67  |   0.00  |   0.00  |   16.67
             | 100.00  |   0.00  |   0.00  |
             |  33.33  |   0.00  |   0.00  |
-------------+---------+---------+---------+
NEUTRAL      |      0  |      0  |      1  |       1
             |   0.00  |   0.00  |  16.67  |   16.67
             |   0.00  |   0.00  | 100.00  |
             |   0.00  |   0.00  |  50.00  |
-------------+---------+---------+---------+
AGREE        |      1  |      1  |      1  |       3
             |  16.67  |  16.67  |  16.67  |   50.00
             |  33.33  |  33.33  |  33.33  |
             |  33.33  | 100.00  |  50.00  |
-------------+---------+---------+---------+
TOTAL               3         1         2         6
                50.00     16.67     33.33    100.00
```

## D. "CHECK ALL THAT APPLY" QUESTIONS

A common problem in questionnaire analysis is "check all that apply" questions. For example, suppose we had a question asking respondents which course or courses they were interested in taking. It might be written:

Which course or courses would you like to see offered next semester?

(check ALL that apply)

```
__1.Micro-computers      __4.Job Control Language
__2.Intro to SAS         __5.FORTRAN
__3.Advanced SAS         __6.PASCAL
```

As far as our analysis is concerned, this is not one question with up to six answers, but six yes/no questions. Each course offering would be treated as a variable with values of YES or NO (coded as 1 or 0, for example). Our questionnaire would be easier to analyze if it were arranged like this:

Please indicate which of the following courses you would like to see offered next semester:

```
                                          | For office
                     (1)yes   (0)no       | use only
a)Micro-computers      __       __        |   __
b)Intro to SAS         __       __        |   __
c)Advanced SAS         __       __        |   __
d)Job Control Language __       __        |   __
e)FORTRAN              __       __        |   __
f)PASCAL               __       __        |   __
```

Our INPUT statement would have six variables (COURSE1-COURSE6 for instance), each with a value of 1 or 0. A format matching 1=yes and 0=no would add readability to the final analysis.

This approach works well when there is a limited number of choices. However, when we are choosing several items from a large number of possible choices, this approach becomes impractical since we need a variable for every possible choice. A common example in the medical field would be the variable "diagnosis" on a patient record. We might have a list of hundreds or even thousands of diagnosis codes and want to consider a maximum of 2 or 3 diagnoses for each patient. Our approach in this case would be to ask for up to 3 diagnoses per

patient, using a diagnosis code from a list of standard-
ized codes. Our form might look like this:

Enter up to 3 diagnosis codes for the patient.

```
    diagnosis 1 _____ |  |__|__|__|
                                     |
    diagnosis 2 _____ |  |__|__|__|
                                     |
    diagnosis 3 _____ |  |__|__|__|
```

Our INPUT statement would be straightforward:

```
DATA DIAG1;
INPUT ID 1-3 . . . DX1 20-22 DX2 23-25 DX3 26-28;
```

Suppose we had the following data:

```
OBS     ID     DX1     DX2     DX3
 1       1      3       4       .
 2       2      1       3       7
 3       3      5       .       .
```

Notice that one patient could have a certain diagnosis code as his first diagnosis while another patient
might have the same code as his second or third diagnosis. If we want a frequency distribution of diagnosis
codes, what can we do? We could try this:

```
PROC FREQ;
   TABLES DX1-DX3;
```

but we would have to add the frequencies from three
tables to compute the frequency for each diagnosis code.
A better approach would be to create a separate data set
that was structured differently. Our goal would be a
data set like this:

```
OBS      ID      DX
 1        1       3
 2        1       4
 3        2       1
 4        2       3
 5        2       7
 6        3       5
```

More details for restructuring data sets can be found in Chapter 7; however, a brief explanation will be given here. The program statements to create the above data set are shown below:

```
1   DATA DIAG2;
2   SET DIAG1;
3   DX=DX1;
4   IF DX NE . THEN OUTPUT;
5   DX=DX2;
6   IF DX NE . THEN OUTPUT;
7   DX=DX3;
8   IF DX NE . THEN OUTPUT;
9   KEEP ID DX;
```

Each observation from our original data set (DIAG1) will create up to 3 observations in our new data set (DIAG2). The SET statement (line 2) reads an observation from our original data set (e.g., ID=1 DX1=3 DX2=3 DX3=.). We then create a new variable, DX, and set it equal to DX1. Line 4 will write an observation to our new data set (DIAG2) as long as the diagnosis is not missing. Each OUTPUT statement creates a new observation in our DIAG2 data set. A simple PROC FREQ with DX as the TABLE variable will now produce the desired diagnosis frequencies.

This program can be made more compact by using an ARRAY and a DO loop.

```
DATA DIAG2;
SET DIAG1;
ARRAY D{*} DX1-DX3;
DO I = 1 TO 3;
   DX=D{I};
   IF D{I} NE . THEN OUTPUT;
   END;
KEEP ID DX;
```

An explanation of ARRAY's can be found in Reference Chapter 1.

## E. LONGITUDINAL DATA

There is a type of data, often referred to as longitudinal data, that needs special attention. Longitudinal data are data that are collected on a group of subjects over time. Caution: the remainder of this chapter is difficult and may be hazardous to your health!

To examine the special techniques needed to analyze longitudinal data, let's follow a simple example. Suppose we are collecting data on a group of patients. (The same scheme would be applicable to periodic data in business or data in the social sciences with repeated measures.) Each time they come in for a visit, we will fill out an encounter form. The data items we collect will be

```
PATIENT ID
DATE OF VISIT (Month Day Year)
HEART RATE
SYSTOLIC BLOOD PRESSURE
DIASTOLIC BLOOD PRESSURE
DIAGNOSIS CODE
DOCTOR FEE
LAB FEE
```

Now, suppose each patient comes in a maximum of 4 times a year. One way to arrange our SAS data set is like this (each visit will be on a separate card):

```
DATA PATIENTS;
INPUT #1 ID1 1-3 DATE1 MMDDYY6.  HR1 10-12 SBP1  13-15 DBP1 16-18
         DX1 19-21 DOCFEE1 22-25 LABFEE1 26-29
      #2 ID2 1-3 DATE2 MMDDYY6.  HR2 10-12 SBP2 13-15 DBP2 16-18
         DX2 19-21 DOCFEE2 22-25 LABFEE2 26-29
      #3 ID3 1-3 DATE3 MMDDYY6.  HR3 10-12 SBP3 13-15 DBP3 16-18
         DX3 19-21 DOCFEE3 22-25 LABFEE3 26-29
      #4 ID4 1-3 DATE4 MMDDYY6.  HR4 10-12 SBP4 13-15 DBP4 16-18
         DX4 19-21 DOCFEE4 22-25 LABFEE4 26-29;
FORMAT DATE1-DATE4 MMDDYY8.;
CARDS;
0071021830701200800140040015o
0071201830721300900200050020o
007
007
0090903830661100701370030000o
009
009
009
0050705830741400820130090000o
0050115820801800960140200150o
0050618820701700840140080040o
0050703830641400840140080020o
```

The number signs (#) in the INPUT statement signify multiple cards per subject. The designation MMDDYY6. following the date variables on the INPUT statement indicates that the date data are in **month day year** form and that they occupy six columns. Dates read in this fashion allow us to do arithmetical computations with date data. (See the SAS User's Guide for more details.) We also included an **output** format for our dates with a **FORMAT** statement. This FORMAT statement uses the same syntax as the earlier examples in this chapter where we created our own formats. The output format MMDDYY8. specifies that the date variables be printed in month/day/year form.

With this method of one card per patient visit, we would need to insert BLANK lines of data for any patient who had less than four visits, to fill out 4 lines per subject. If we wanted to compute within-subject means, we would continue (before the CARDS statement)

```
AVEHR = MEAN (OF HR1-HR4);
AVESBP = MEAN (OF SBP1-SBP4);
AVEDBP = MEAN (OF DBP1-DBP4);
            etc.
```

where "MEAN" is one of the SAS built-in functions that will compute the mean of all the variables listed in parentheses. (Note: if any of the variables listed as arguments of the MEAN function have missing values, the result will be the mean of the **nonmissing** values.) A much better approach would be to treat **EACH VISIT AS A SEPARATE OBSERVATION.** Our program would then look like this:

```
DATA PATIENTS;
INPUT ID 1-3 DATE MMDDYY6. HR 10-12 SBP 13-15
      DBP 16-18 DX 19-21 DOCFEE 22-25 LABFEE 26-29;
FORMAT DATE MMDDYY8.;
CARDS;
```

Now, we need to include only as many lines of data as there are patient visits; blank lines to fill out 4 lines per subject are not needed. Our variable names are also simpler since we do not need to keep track of HR1, HR2, etc. How do we do analyses on this data set? A simple PROC MEANS on a variable such as HR,SBP, or DOCFEE will not be particularly useful since we are averaging 1 to 4 values per patient together. Perhaps the average of DOCFEE would be useful since it represents the average doctor fee **per** PATIENT VISIT, but statistics for heart rate or blood pressure would be a weighted average, the weight depending on how many visits we had from each patient. How do we compute the average heart rate or blood pressure PER PATIENT? The key is to use **ID as a BY variable.**

Here is our program (with sample data):

```
DATA PATIENTS;
INPUT ID 1-3 DATE MMDDYY6. HR 10-12 SBP 13-15
     DBP 16-18 DX 19-21 DOCFEE 22-25 LABFEE 26-29;
FORMAT DATE MMDDYY8.;
CARDS;
0071021830701200800140040010150
0071201830721300900200050010200
0090903830661100701370030010000
0050705830741400820130090010000
0050115820801800960140200015000
0050618820701700840140080010400
0050703830641400840140080010200
PROC SORT;
   BY ID DATE;
PROC MEANS NOPRINT;
   BY ID;
   VAR HR -- DBP DOCFEE LABFEE;
   OUTPUT OUT=STATS MEAN=HR SBP DBP DOCFEE LABFEE;
```

What have we done here? The PROC SORT statement assures us that the data set is in patient-date order. We can now use ID as a BY variable with PROC MEANS. The result is the mean HR, SBP, etc. per patient, which is placed in a new data set STATS, created with the **OUTPUT** statement of PROC MEANS. (See section G of Reference Chapter 2, **Procedures that Output Data Sets.**) Each variable listed after "MEAN=" in the OUTPUT statement will be the **mean** of the variables listed in the **VAR** statement, in the **order** they appear. Thus, HR in the data set STATS will be the mean HR in the data set PATIENTS. The variable names you choose following "MEAN=" are arbitrary; they do **not** have to be the same as the names in the VAR list. It is the **order** of the variable names corresponding to the **order** of the VAR list that is important. Because of the **BY** variable, the mean HR etc. will be computed for **each subject.** The **NOPRINT** option simply says to create the new data set **STATS** without printing the PROC MEANS statistics in our output. (This output, had we omitted the NOPRINT option, would be the mean HR etc. for EACH patient.) In this example, the data set STATS would look like this (we can always test this with a PROC PRINT statement).

| OBS | ID | HR | SBP | DBP | DOCFEE | LABFEE |
|-----|----|------|--------|------|--------|--------|
| 1 | 5 | 71.33 | 157.50 | 86.5 | 112.5 | 525. |
| 2 | 7 | 71.00 | 125.00 | 85.0 | 45.0 | 175. |
| 3 | 9 | 66.00 | 110.00 | 70.0 | 30.0 | 0. |

This data set contains the mean HR, SBP, etc. PER patient. We could analyze this data set with additional SAS procedures to investigate relationships between variables or compute descriptive statistics where each data value corresponds to a single value (the MEAN) from each patient.

## F. MOST RECENT (OR LAST) VISIT PER PATIENT

What if we want to analyze the most recent visit for each patient? The data set PATIENTS in our previous example was sorted in patient-date order. The most recent visit would be the last observation for each patient ID. We can extract these observations with the following SAS program:

```
 1. DATA PATIENTS;
 2. INPUT ID 1-3 DATE MMDDYY6. HR 10-12 SBP 13-15
      DBP 16-18 DX 19-21 DOCFEE 22-25 LABFEE 26-29;
 3. CARDS;
 4. FORMAT DATE MMDDYY8.;
    (DATA CARDS)
 5. PROC SORT;
 6.    BY ID DATE;
 7. DATA RECENT;
 8. SET PATIENTS;
 9.    BY ID;
10. IF LAST.ID;
```

We have introduced several new features of SAS software: Line 7 creates a new data set called RECENT. The SET statement in line 8 acts like an INPUT statement except that observations are read one by one from the SAS data set PATIENTS instead of our original (card) data.

We are permitted to use a BY variable following our SET statement providing our data set has been previously sorted by the same variable (it has). The effect of adding the BY statement is to allow us to use the special **FIRST.** and **LAST.** internal SAS variables. Since ID was our BY variable, FIRST.ID will be a logical variable (i.e., true or false: 1 or 0) that will be part of each observation as it is being processed but will not remain with the final data set RECENT. **FIRST.ID** will be **TRUE** (or 1) whenever we are reading a **new ID**; **LAST.ID** will be **TRUE** whenever we are reading the **last** observation for a given ID. To clarify this, here are our observations and the value of FIRST.ID and LAST.ID in each case:

| OBS | ID | DATE | HR | SBP | DBP | DX | DOCFEE | LABFEE | FIRST.ID | LAST.ID |
|-----|----|------|----|-----|-----|----|--------|--------|----------|---------|
| 1 | 5 | 01/15/82 | 80 | 180 | 96 | 14 | 200 | 1500 | 1 | 0 |
| 2 | 5 | 06/18/82 | 70 | 170 | 84 | 14 | 80 | 400 | 0 | 1 |
| 3 | 5 | 07/03/83 | 64 | 140 | 84 | 14 | 80 | 200 | 0 | 1 |
| 4 | 5 | 07/05/83 | 74 | 140 | 82 | 13 | 90 | 0 | 0 | 1 |
| 5 | 7 | 10/21/83 | 70 | 120 | 80 | 14 | 40 | 150 | 1 | 0 |
| 6 | 7 | 12/01/83 | 72 | 130 | 90 | 20 | 50 | 200 | 0 | 1 |
| 7 | 9 | 09/03/83 | 66 | 110 | 70 | 137 | 30 | 0 | 1 | 0 |

By adding the IF statement in line 10, we can select the last visit for each patient (in this case, observations 4, 6, and 7).

### G. COMPUTING FREQUENCIES ON LONGITUDINAL DATA SETS

To compute frequencies for our diagnoses, we would use PROC FREQ on our original data set (PATIENTS). We would write

```
PROC FREQ DATA=PATIENTS ORDER=FREQ;
   TITLE 'DIAGNOSES IN DECREASING FREQUENCY ORDER';
   TABLES DX;
```

Notice we use the DATA= option on the PROC FREQ statement to make sure we were counting frequencies from our original data set. The ORDER= option allows us to control the order of the categories in a PROC FREQ output. Normally, the diagnosis categories would be listed in numerical order. The ORDER=FREQ option lists the diagnoses in frequency order from the most common diagnosis to the least. While we are on the subject, another useful ORDER= option is ORDER=FORMATTED. This will order the diagnoses alphabetically by the diagnosis formats (if we had them). Remember that executing a PROC FREQ procedure on our original data set (the one with 1 to 4 observations per patient) will have the effect of counting the number of times each diagnosis was made. That is, if a patient came in for three visits, each with the same diagnosis, we would add 3 to the frequency count for that diagnosis. If, for some reason, we want to count a diagnosis only once for a given patient, even if this diagnosis is made on subsequent visits, we can **sort our data set by ID and DX.** We then have a data set such as the following:

```
ID      DX    FIRST.ID   FIRST.DX
-------------------------------
 5      13       1          1
 5      14       0          1
 5      14       0          0
 5      14       0          0
 7      14       1          1
 7      20       0          1
 9     137       1          1
```

If we now use the logical FIRST.DX and FIRST.ID variables, we can accomplish our goal of counting a diagnosis only once for a given patient. The logical variable FIRST.DX will be **true** each time a new ID-diagnosis combination is encountered. The data set and procedure would look as follows (assume we have previously sorted by ID and DX):

```
DATA DIAG;
SET PATIENTS;
   BY ID DX;
IF FIRST.DX;
PROC FREQ ORDER=FREQ;
   TABLES DX;
```

# Chapter 4    Correlation and Regression

## A. CORRELATION

A common statistic to show the strength of a relationship that exists between two continuous variables is
called the **Pearson correlation coefficient** or just correlation coefficient (there are other types of correlation coefficients). The correlation coefficient is a
number that ranges from -1 to +1. A positive correlation
means that as values of one variable increase, values of
the other variable will also tend to increase. A small
or zero correlation coefficient would tell us that the
two variables were unrelated. Finally, a negative correlation coefficient shows an inverse relationship
between the variables: as one goes up, the other goes
down. Before we discuss correlations any further, we
should calculate the correlation coefficient between
height and weight in one of our earlier examples. The
entire program would look as follows:

```
DATA HTWT;
INPUT SEX $ HEIGHT WEIGHT;
CARDS;
M 68 155
F 61  99
F 63 115
M 70 205
M 69 170
F 65 125
M 72 220
PROC CORR;
    VAR HEIGHT WEIGHT;
```

The general form of PROC CORR is

```
PROC CORR;
    VAR  variables;
```

where "variables" is replaced with a list of variable
names, separated by spaces. This procedure found a correlation coefficient of 0.97165 between the variables
HEIGHT and WEIGHT.

Many people will ask, "How large a correlation coefficient do I need to show that two variables are correlated?" Each time PROC CORR prints a correlation coefficient, it also prints a probability associated with the coefficient. This number gives the probability of obtaining a sample correlation coefficient as large as or larger than the one obtained, **by chance alone** (that is, when the variables in question actually have zero correlation). We see below the output from PROC CORR:

```
VARIABLE  N        MEAN      STD DEV        SUM      MINIMUM      MAXIMUM

HEIGHT    7    66.85714     3.976119    468.0000    61.00000     72.00000
WEIGHT    7   155.57143    45.796132   1089.0000    99.00000    220.00000

CORRELATION COEFFICIENTS / PROB > |R| UNDER H0:RHO=0 / N = 7
               HEIGHT     WEIGHT

HEIGHT     1.00000    0.97165
           0.0000     0.0003
WEIGHT     0.97165    1.00000
           0.0003     0.0000
```

The significance of a correlation coefficient is a function of the magnitude of the correlation and the sample size. With a large number of data points, even a small correlation coefficient can be significant. For example, with 10 data points, a correlation coefficient of .63 or larger is significant; with 100 data points, a correlation of .195 would be significant (at the .05 level). Note that a negative correlation shows an equally strong relationship as a positive correlation (although the relationship is inverse). A correlation of -.40 is just as strong as one of +.40. It is important to remember that correlation indicates only the strength of a relationship--**it does not imply causality**. For example, we would probably find a high positive correlation between the number of hospitals in each of the 50 states versus the number of household pets in each state. Does this mean that pets make people sick and therefore make more hospitals necessary? Doubtful. The most plausible explanation would be that both variables (number of pets and number of hospitals) are related to population.

Being **SIGNIFICANT** is not the same as being **IMPORTANT** or **STRONG**. That is, knowing the significance of the correlation coefficient does not tell us very much. Once we know that our correlation coefficient is significantly different from zero, we need to look further in interpreting the importance of this correlation. Let us digress a moment to ask what we mean by the significance of a correlation coefficient. Suppose we had a population of x and y values in which the correlation was zero. We could imagine a plot of this population as shown below:

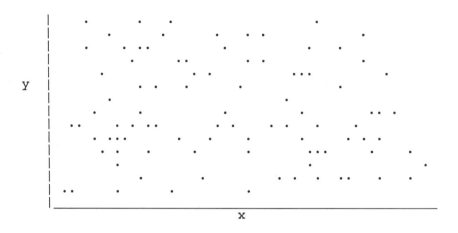

Suppose further that we choose a small sample of 10 points from this population. In the plot below, the o's represent the x,y pairs we choose "at random" from our population.

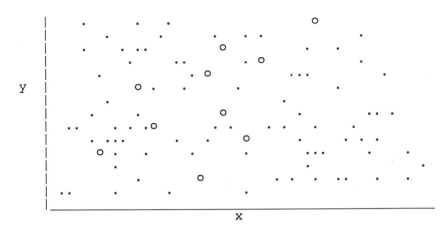

Notice that "by chance alone," we can wind up with a nonzero correlation coefficient. As our sample size increases, it becomes more and more **unlikely** to choose a sample at random that would have a large correlation coefficient. The lesson here is: Whenever we compute a correlation coefficient, we must find the probability of obtaining a correlation this large or larger by chance alone.

One of the best ways of interpreting a correlation coefficient (r) is to look at the square of the coefficient (r squared). R squared can be interpreted as the proportion of variance in one of the variables that can be explained by variation in the other variable. As an example, our height/weight correlation is .97. Thus, r squared is .94. We can say that 94% of the variation in weights can be explained by variation in height (or vice versa). Also, (1 - .94) or 6% of the variance of weight is due to factors other than height variation. With this interpretation in mind, if we examine a correlation coefficient of .4, we see that only .16 or 16% of the variance of one variable is explained by variation in the other.

Another consideration in the interpretation of correlation coefficients is this: **Be sure to look at a scatter plot of the data** (using PROC PLOT). It often turns out that one or two extreme data points can cause the correlation coefficient to be much larger than expected.

An important assumption concerning a correlation coefficient is that each pair of x,y data points is independent of any other pair. That is, each pair of points has to come from a separate subject. Suppose we had 20 subjects and we measured two variables (say blood pressure and heart rate) on each of the 20 subjects. We could then calculate a correlation coefficient using 20 pairs of data points. Suppose instead that we made 10 measurements of blood pressure and heart rate on **each** subject. We cannot compute a valid correlation coefficient using all 200 data points (10 points for each of 20 subjects) because the 10 points for each subject are not independent. The best method would be to compute the mean blood pressure and heart rate for each subject and use the mean values to compute the correlation coefficient. Having done so, we'll keep in mind that we correlated **mean** values when we are interpreting the results.

## B. LINEAR REGRESSION

Given a person's height, what would be the predicted weight? How can we best define the relationship between height and weight? By looking at the graph below we see that the relationship is approximately linear. That is, we can imagine drawing a straight line on the graph with most of the data points being only a short distance from the line. **The vertical distance from each data point to this line is called a residual.**

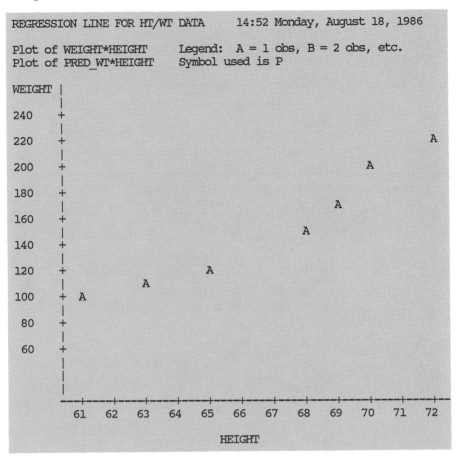

```
REGRESSION LINE FOR HT/WT DATA     14:52 Monday, August 18, 1986

Plot of WEIGHT*HEIGHT      Legend:  A = 1 obs, B = 2 obs, etc.
Plot of PRED_WT*HEIGHT     Symbol used is P

WEIGHT |
       |
  240  +
       |
  220  +                                                          A
       |
  200  +                                             A
       |
  180  +
       |                                      A
  160  +
       |                               A
  140  +
       |
  120  +                 A
       |          A
  100  + A
       |
   80  +
       |
   60  +
       |
       |
       |
       ---+----+----+----+----+----+----+----+----+----+----+----+--
          61   62   63   64   65   66   67   68   69   70   71   72
                                HEIGHT
```

How do we determine the **"best"** straight line to fit our height/weight data? The method of **least squares** is commonly used. This method finds the line (called the **regression line**) that minimizes the **sum** of the **squared residuals**. A residual is the difference between a subject's predicted score and his/her actual score.

Using our height/weight program, we add the following PROC to give us the equation for the regression line:

```
PROC REG;
    TITLE 'REGRESSION LINE FOR HT/WT DATA';
    MODEL WEIGHT=HEIGHT;
```

PROC REG (short for regression) has the general form

```
PROC REG;
    MODEL  dependent variable(s) = independent variable;
```

The output from the above procedure is shown below:

```
REGRESSION LINE FOR HT/WT DATA

DEP VARIABLE: WEIGHT
                   SUM OF        MEAN
SOURCE      DF     SQUARES       SQUARE      F VALUE      PROB>F
MODEL       1      11880.327     11880.327   84.451       0.0003
ERROR       5      703.387       140.677
C TOTAL     6      12583.714
    ROOT MSE       11.860751     R-SQUARE    0.9441
    DEP MEAN       155.571       ADJ R-SQ    0.9329
    C.V.           7.62399


                   PARAMETER     STANDARD    T FOR H0:
VARIABLE    DF     ESTIMATE      ERROR       PARAMETER=0  PROB > |T|

INTERCEP    1      -592.645      81.542175   -7.268       0.0008
HEIGHT      1      11.191265     1.217803    9.190        0.0003
```

Look  first  at the bottom few lines.  There is  an
estimate  for  two  parameters,  intercept  and  height.
Remember  that the general equation for a straight  line
can be written as

$$y = mx + b$$

where

    m = slope
    b = intercept.

(Statisticians  often use **a** or **b**$_0$  for the intercept and
**b** or **b**$_1$ for the slope.)

We can write the equation for the "best" straight line defined by our height/weight data as

$$WEIGHT = (11.19)(HEIGHT) + (-592.64)$$
$$= (11.19)(HEIGHT) - 592.64$$

Given any height, we can now predict the weight. For example, the predicted weight of a 70-inch-tall person would be:

$$WEIGHT = (11.19)(70) - 592.64 = 190.66 \text{ LBS.}$$

To the right of the parameter estimates, we see columns labeled STANDARD ERROR, T FOR H0:PARAMETER=0, and PROB > |T|. The T values and the associated probabilities (PROB >|T|) test the hypothesis that the parameter is actually zero. That is, if the true slope and intercept were zero, what would the probability be of obtaining, by chance alone, a value as large as or larger than the one actually obtained? The STANDARD ERROR can be thought of in the same way as the standard error of the mean. It reflects the accuracy with which we know the true slope and intercept.

In our case, the slope is 11.19 and the standard error of the slope is 1.22. We can therefore form a 95% confidence interval for the slope by taking two (approximately) standard errors above and below the mean. The 95% confidence interval for our slope is 8.75 to 13.63. Actually, since the number of points in our example is small (n=7) we really should go to a t-table to find the number of standard errors above and below the mean for a 95% confidence interval. (This is true when n is less than 30.) Going to a t-table, we look under degrees of freedom (df) equal to n-1 and level of significance (two-tail) equal to .05. The value of t for df=6, and p=.05 is 2.45. Our 95% confidence interval is then 11.19 plus or minus (2.45)(1.22)=2.99.

Inspecting the indented portion of the output, we see values for ROOT MSE, R-SQUARE, DEP MEAN, ADJ R-SQ and C.V. ROOT MSE is the square root of the error variance. That is, it is the standard deviation of the residuals. R-SQUARE is the square of the generalized correlation coefficient. Since we have only one independent variable (HEIGHT), R-SQUARE is the square of the Pearson correlation coefficient between HEIGHT and WEIGHT and, as we discussed in the previous section, the square of the correlation coefficient tells us how much variation in the dependent variable can be accounted for by variation of the independent variable. When there is more than one independent variable, R-SQUARE will reflect the variation in the dependent variable accounted for by a linear combination of all the independent variables (see Chapter 8 for a more complete discussion). The DEP MEAN is the mean of the dependent variable: WEIGHT in this case. C.V. is the coefficient of variation discussed in Chapter 2. Finally, ADJ R-SQ is the squared correlation coefficient corrected for the number of independent variables in the equation. This adjustment has the effect of **decreasing** the value of $R^2$. The difference is typically small but becomes larger and more important when dealing with multiple independent variables (see Chapter 8).

Finally, the top portion of the output shows sources of variation. Each weight differs from the mean weight for two reasons: (1) because each person has a different height and (2) because each person is above or below the predicted weight for his or her height.

The **total sum of squares** is the sum of squared deviations of each person's weight from the mean weight of the group. This total sum of squares (SS) can be divided into the two portions mentioned above. One portion, called the **ERROR SUM OF SQUARES** in the output, reflects the deviation of each weight from the PREDICTED weight. The other portion reflects deviations between the PREDICTED values and the MEAN. This is called the SUM OF SQUARES due to the **MODEL** in the output. The column labeled **MEAN SQUARE** is the **SUM OF SQUARES** divided by the **degrees of freedom.** In our case, there are seven data points. The **total** degrees of freedom is equal to n-1 or 6. The **model** has two parameters and has 1 df. The error degrees of freedom is the **total** df minus the **model** df which gives us 5. We can think of the **MEAN  SQUARE** as the respective variance (square of the standard deviation) of each of these two portions of variation. Our intuition tells us that if the deviations about the regression line are small (ERROR MEAN SQUARE) compared to the deviation between the predicted values and the mean (MODEL MEAN SQUARE) then we have a good regression line. To compare these mean squares, the ratio

```
        MEAN SQUARE MODEL
F =  -------------------
        MEAN SQUARE ERROR
```

is formed. The larger this ratio, the better the fit (for a given number of data points). The program prints this F value and the probability of obtaining an F this large or larger by chance alone. In the case where there is only one independent variable, the probability of the F statistic is exactly the same as the probability associated with testing for the significance of the correlation coefficient. If this probability is "large" (over .05) then our linear model is not doing a good job of describing the relationship between the variables.

To plot out height/weight data, we can use PROC PLOT as follows:

```
PROC PLOT;
   PLOT WEIGHT*HEIGHT;
```

However, we would have to draw in the regression line by hand, using the slope and intercept from PROC REG. It would be desirable to have SAS software plot the regression line (actually, the regression predicted values) for us. To do this, we have to request PROC REG to **output** a data set containing the original height/weight data as well as a new variable representing the **predicted** weight. This is accomplished with an **OUTPUT** statement following PROC REG. The form is

```
OUTPUT OUT=data set name PREDICTED=variable name;
```

For example

```
PROC REG;
   MODEL WEIGHT=HEIGHT;
   OUTPUT OUT=REGOUT PREDICTED=PRED_WT;
```

will produce a new SAS data set called REGOUT (use any valid SAS data set name). This data set will contain **all** the original variables in the data set as well as the computed variable, PRED_WT, which is the predicted weight for each height. (Note: the underscore in the variable name PRED_WT is a valid character in any SAS variable name.) It was requested by the keyword **PREDICTED=** in the OUTPUT statement. To be sure this is clear, let's follow PROC REG with a PROC PRINT statement to allow us to see the contents of the newly created data set, REGOUT. Remember, **all** PROC's will perform calculations on the **most recently created data set** (unless we specify a different data set with the DATA= option following the PROC). Therefore, since REGOUT was just created by use of an OUTPUT statement following PROC REG, the PROC's following it will operate on REGOUT and **not** HTWT (the original data set). Below is the output from this PROC PRINT:

```
OBS SEX HEIGHT WEIGHT PRED_WT

  1   M    68     155   168.361
  2   F    61      99    90.023
  3   F    63     115   112.405
  4   M    70     205   190.744
  5   M    69     170   179.553
  6   F    65     125   134.788
  7   M    72     220   213.127
```

Now that we have a data set with the variables HEIGHT, WEIGHT, and PRED_WT, we can produce a plot of the orginal data as well as the predicted values (i.e., the points on the regression line). Here are the SAS statements to compute the regression line and to plot the results:

```
PLOT REG;
   MODEL WEIGHT=HEIGHT;
   OUTPUT OUT=REGOUT PREDICTED=PRED_WT;
PROC PLOT;
   PLOT WEIGHT*HEIGHT PRED_WT*HEIGHT='P' / OVERLAY;
```

The PLOT request above asks for two plots: WEIGHT versus HEIGHT (the original data points) and PRED_WT versus HEIGHT (the points on the regression line). The ='P' specification following "PRED_WT*HEIGHT" requests that the letter "P" be used as the plotting symbol for this plot. This is to distinguish the original data points (plotted as A's, B's, etc.) from the regression line points. We do this since all data points will be on **one** graph, thanks to our **OVERLAY** option. As you probably guessed, OVERLAY causes **all** plot requests to be printed on a single set of axes instead of a separate plot for each x,y specification. The resulting graph is displayed below:

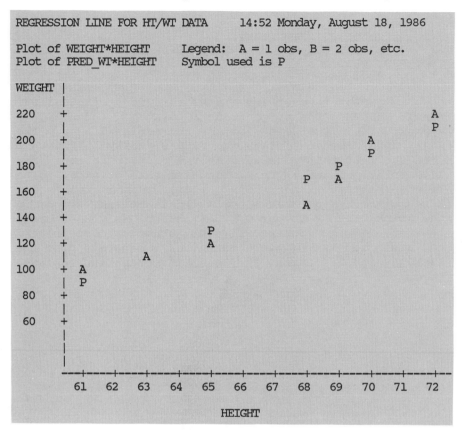

```
REGRESSION LINE FOR HT/WT DATA          14:52 Monday, August 18, 1986

Plot of WEIGHT*HEIGHT      Legend:  A = 1 obs, B = 2 obs, etc.
Plot of PRED_WT*HEIGHT     Symbol used is P

WEIGHT |
       |
  220  +                                                              A
       |                                                              P
  200  +                                                    A
       |                                                    P
  180  +                                          P
       |                                     P    A
  160  +
       |                                A
  140  +
       |                  P
  120  +                  A
       |           A
  100  +  A
       |  P
   80  +
       |
   60  +
       |
       |
       |
       --+----+----+----+----+----+----+----+----+----+----+----+--
         61   62   63   64   65   66   67   68   69   70   71   72
                                 HEIGHT
```

Before we leave this section, we should mention that SAS software has the ability to **OUTPUT** several computed values other than the PREDICTED value of the dependent variable. The most useful statistics besides the predicted values are

RESIDUAL=var name               The named variable will be the
                                RESIDUAL (i.e., difference be-
                                tween    actual   and   predicted
                                values for each observation).

L95=var name                    The  Lower or **Upper** 95%  confi-
U95=var name                    dence  limit for an  individual
                                predicted value. (i.e., 95% of
                                the dependent data values would
                                be expected to be between these
                                two limits).

L95M=var name                   The  Lower or **Upper** 95%  confi-
U95M=var name                   dence limit for the **mean** of the
                                dependent variable for a  given
                                value  of the independent vari-
                                able.

To demonstrate some of these options in action,  we
will produce the following:

(a)  A plot of the original data,  predicted values,
and  the upper and lower 95%  confidence limits  for
the mean weight.

(b) A plot of residuals versus HEIGHT. Note: If this
plot  shows  some  systematic pattern (rather  than
random  points),   one  could  try  to  improve  the
original model.

Here is the program to accomplish the above requests:

```
PROC REG;
   MODEL WEIGHT=HEIGHT;
   OUTPUT OUT=REGOUT
          PREDICTED=PRED_WT           RESIDUAL=RES_WT
          L95M=L_WT                   U95M=U_WT;
PROC PLOT;
   PLOT WEIGHT*HEIGHT PRED_WT*HEIGHT='P'
        L_WT*HEIGHT='L' U_WT*HEIGHT='U' / OVERLAY;
PLOT RES_WT*HEIGHT;
```

The two graphs are shown below:

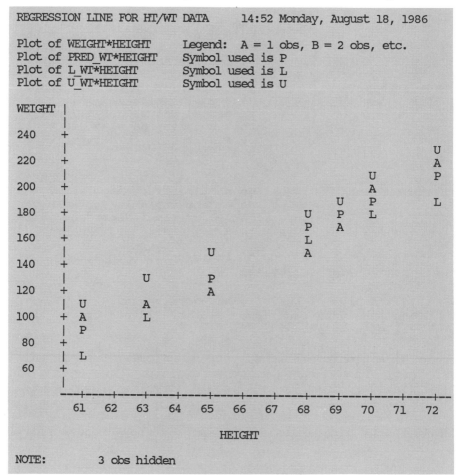

```
REGRESSION LINE FOR HT/WT DATA      14:52 Monday, August 18, 1986

Plot of WEIGHT*HEIGHT        Legend:  A = 1 obs, B = 2 obs, etc.
Plot of PRED_WT*HEIGHT       Symbol used is P
Plot of L_WT*HEIGHT          Symbol used is L
Plot of U_WT*HEIGHT          Symbol used is U

WEIGHT |
       |
  240  +
       |                                                                    U
  220  +                                                                    A
       |                                                    U               P
  200  +                                                    A
       |                                        U           P               L
  180  +                            U           P           L
       |                            P           A
  160  +                            L
       |                U           A
  140  +
       |       U        P
  120  +                A
       | U     A
  100  + A     L
       | P
   80  +
       | L
   60  +
       |
       ---+----+----+----+----+----+----+----+----+----+----+----+--
          61   62   63   64   65   66   67   68   69   70   71   72
                                   HEIGHT

NOTE:       3 obs hidden
```

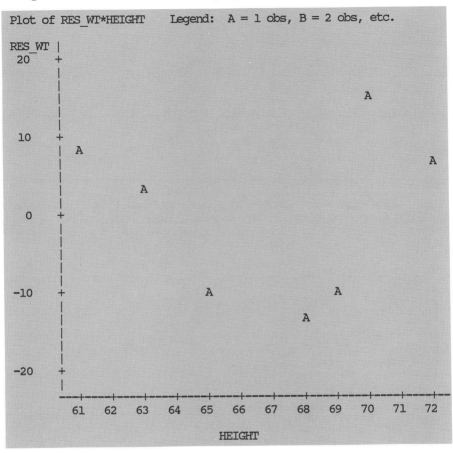

Plot of RES_WT*HEIGHT    Legend: A = 1 obs, B = 2 obs, etc.

## C. TRANSFORMING DATA

Another regression example is given here to demonstrate some additional steps that may be necessary when doing regression. Shown below are data that were collected on 10 people:

| Subject | Drug dose | Heart rate |
| --- | --- | --- |
| 1 | 2 | 60 |
| 2 | 2 | 58 |
| 3 | 4 | 63 |
| 4 | 4 | 62 |
| 5 | 8 | 67 |
| 6 | 8 | 65 |
| 7 | 16 | 70 |
| 8 | 16 | 70 |
| 9 | 32 | 74 |
| 10 | 32 | 73 |

Let's write a SAS program to define this collection of data and to plot drug dose by heart rate.

```
DATA;
INPUT DOSE HR;
CARDS;
2 60
   .
   .
   .
32 73
 PROC PLOT;
    PLOT HR*DOSE;
 PROC REG;
    MODEL HR = DOSE;
```

The resulting graph and the PROC REG output are shown below:

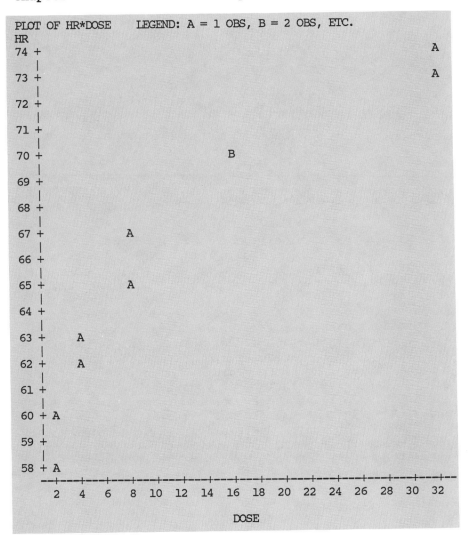

```
PLOT OF HR*DOSE     LEGEND: A = 1 OBS, B = 2 OBS, ETC.
HR
74 +                                                            A
   |
73 +                                                            A
   |
72 +
   |
71 +
   |
70 +                             B
   |
69 +
   |
68 +
   |
67 +           A
   |
66 +
   |
65 +           A
   |
64 +
   |
63 +      A
   |
62 +      A
   |
61 +
   |
60 + A
   |
59 +
   |
58 + A
   --+---+---+---+---+---+---+---+---+---+---+---+---+---+---+--
      2   4   6   8  10  12  14  16  18  20  22  24  26  28  30  32
                                 DOSE
```

```
DEP VARIABLE: HR
                    SUM OF          MEAN
SOURCE      DF      SQUARES         SQUARE      F VALUE      PROB>F
MODEL       1       233.484         233.484     49.006       0.0001
ERROR       8       38.115591       4.764449
C TOTAL     9       271.600
    ROOT MSE        2.182762        R-SQUARE    0.8597
    DEP MEAN        66.200000       ADJ R-SQ    0.8421
    C.V.            3.297223


                    PARAMETER       STANDARD    T FOR H0:
VARIABLE    DF      ESTIMATE        ERROR       PARAMETER=0   PROB > |T|

INTERCEP    1       60.708333       1.044918    58.099        0.0001
DOSE        1       0.442876        0.063264    7.000         0.0001
```

Either by clinical judgment or by careful inspection of the graph, we decide that the relationship is not linear. We see an approximately equal increase in heart rate each time the dose is doubled. Therefore, if we plot log dose against heart rate we can expect a linear relationship. SAS software has a number of built-in functions such as logarithms and trigonometric functions. These functions are described in Chapter 6 of the **SAS User's Guide: Basics** Version 5, and in Chapter 4 in the **SAS Language Guide for Personal Computers**, Version 6. We can write mathematical equations to define new variables by placing these statements between the INPUT and CARDS statements. In SAS programs, we represent addition, subtraction, multiplication, and division by the symbols +,-,*, and /, respectively. Exponentiation is written as **. To create a new variable which is the log of dose, we write

```
DATA;
INPUT DOSE HR;
LDOSE = LOG(DOSE);
CARDS;
```

where LOG is a SAS function that yields the natural (base e) logarithm of whatever value is in the parentheses.

We can now plot log dose versus heart rate and compute a new regression line.

```
PROC PLOT;
   PLOT HR*LDOSE;
PROC REG;
   MODEL HR=LDOSE;
```

Output from the above statements is shown on the following pages. Approach transforming variables with caution. Keep in mind that when a variable is transformed, one should not refer to the variable as in the untransformed state. That is, don't refer to the "log of dosage" as "dosage." Some variables are frequently transformed: income, sizes of groups, and magnitudes of earthquakes are usually presented as logs, or in some other transformation.

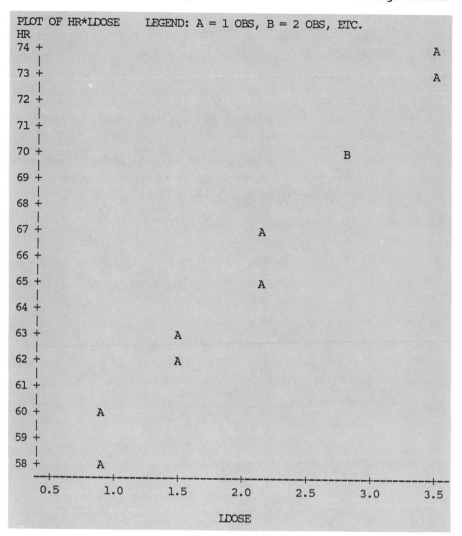

```
PLOT OF HR*LDOSE      LEGEND: A = 1 OBS, B = 2 OBS, ETC.
HR
74 +                                                        A
   |
73 +                                                        A
   |
72 +
   |
71 +
   |
70 +                                              B
   |
69 +
   |
68 +
   |
67 +                            A
   |
66 +
   |
65 +                            A
   |
64 +
   |
63 +                  A
   |
62 +                  A
   |
61 +
   |
60 +        A
   |
59 +
   |
58 +        A
  --+---------+---------+---------+---------+---------+---------+--
    0.5      1.0       1.5       2.0       2.5       3.0       3.5
                              LDOSE
```

## Chapter 4   Correlation and Regression

```
DEP VARIABLE: HR
                    SUM OF          MEAN
SOURCE      DF      SQUARES         SQUARE       F VALUE      PROB>F
MODEL       1       266.450         266.450      413.903      0.0001
ERROR       8       5.150000        0.643750
C TOTAL     9       271.600
       ROOT MSE     0.802340        R-SQUARE     0.9810
       DEP MEAN     66.200000       ADJ R-SQ     0.9787
       C.V.         1.211994

                    PARAMETER       STANDARD     T FOR H0:
VARIABLE    DF      ESTIMATE        ERROR        PARAMETER=0    PROB > |T|

INTERCEP    1       55.250000       0.595032     92.852         0.0001
LDOSE       1       5.265837        0.258832     20.345         0.0001
```

Notice that the data points are now closer to the regression line. The MEAN SQUARE ERROR term is smaller and R SQUARE is larger, confirming our conclusion that dose versus heart rate fits a logarithmic curve better than a linear one.

## EST: TESTING DIFFERENCES BETWEEN TWO MEANS

A common experimental design is to randomly assign subjects to a treatment or a control group and then measure one or more variables that would be hypothesized to be affected by the treatment. In order to determine whether the means of the treatment and control groups are significantly different, we set up what is called a "null hypothesis" ($H_0$). It states that the treatment and control groups would have the same mean if we repeated the experiment a large (infinite) number of times and that the differences in any one trial are attributable to the luck of the draw in assigning subjects to treatment and control groups. The alternative hypotheses ($H_1$) to the null hypothesis are that one particular mean will be greater than the other (called a 1-tailed test) or that the two means will be different, but the researcher cannot say a priori which will be greater (called a 2-tailed test). The researcher can specify either of the two alternative hypotheses. When the data have been collected, a procedure called the t-test is used to determine the probability that the difference in the means that is observed is due to chance. The lower the likelihood that the difference is due to chance, the greater the likelihood that the difference is due to there being real differences in treatment and control. The following example will demonstrate a SAS program that performs a t-test.

Students are randomly assigned to a control or treatment group (where a drug is administered). Their response time to a stimulus is then measured. The times are as follows:

```
        (Response time in milliseconds)
        Control                Treatment
        ------------------------------------------
          90                     100
          93                     103
          87                     104
          89                      99
          90                     102
```

Is the mean of the treatment group significantly different from the mean of the control group? A quick calculation shows the mean of the control group to be 89.8 and the mean of the treatment group to be 101.6. You may recall the discussion in Chapter 2 about the standard error of the mean. When we use a sample to estimate the population mean, the standard error of the mean reflects how accurately we can estimate the population mean. As you will see when we write a SAS program to analyze this experiment, the standard error of the control mean is .97 and the standard error of the treatment mean is .93. Since the means are 11.8 units apart, even if each mean is several standard errors away from its true population mean, they would still be significantly different from each other. There are some assumptions that must be met before we can apply this test. First, the two groups must be independent. This will be ensured with our method of random assignment. Second, the theoretical distribution of sampling means should be normally distributed (this will be ensured if the sample size is sufficiently large) and third, the variances of the two groups should be approximately equal. This last assumption is checked automatically each time a t-test is computed by the SAS system. The SAS t-test output contains t-values and probabilities for both the case of equal group variances and unequal group variances. We will see later how to test whether the data are normally distributed. Finally, when the t-test assumptions are not met, there are other procedures that can be used. These will be demonstrated later in this chapter.

It's time now to write a program to describe our data and to request a t-test. What are our variables?

What information have we collected on each subject? We know which group each person belongs to and what his response time is. The program can thus be written as follows:

```
DATA;
INPUT GROUP $ TIME;
CARDS;
C 80
C 93
C 83
C 89
C 98
T 100
T 103
T 104
T 99
T 102
PROC TTEST;
    CLASS GROUP;
    VAR TIME;
```

PROC PRINT will give us a listing of our original data. It is convenient to include a PROC PRINT with most of our SAS programs so that we will have a listing of our data along with the analyses.

PROC TTEST has two parts. Following the word CLASS is the **independent** variable--the variable that identifies the two groups of subjects. In our case, the variable GROUP has values of C (for the control group) and T (for the treatment group).

The variable or variable list that follows the word VAR identifies the **dependent** variable(s), in our case, TIME. When more than one **dependent** variable is listed, a **separate** t-test is computed for each dependent variable in the list.

Look at the TTEST output below:

```
Ttest Procedure

Variable: TIME

GROUP         N              Mean          Std Dev        Std Error
-------------------------------------------------------------------
C             5         88.60000000      7.30068490      3.26496554
T             5        101.60000000      2.07364414      0.92736185

Variances         T        DF     Prob>|T|
-------------------------------------------
Unequal       -3.8302      4.6     0.0145
Equal         -3.8302      8.0     0.0050

For H0: Variances are equal, F'=   12.40 with 4 and 4 DF
        Prob > F'= 0.0318
```

We see the mean value of TIME for the control and treatment groups, along with the standard deviation and standard error. Below this are two sets of t-values, degrees of freedom, and probabilities. One is valid if we have equal variances, the other if we have unequal variances. You will usually find these values very close unless the variances differ widely. The bottom line gives us the probability that the variances are unequal due to chance. If this probability is small (say less than .05) then we are going to reject the hypothesis that the variances are equal. We then use the t-value and probability labeled UNEQUAL. If the PROB > F' value is greater than .05, use the t-value and probability for equal variances.

In this example, we look at the bottom of the t-test output and see that the F ratio (larger variance divided by the smaller variance) is 12.40. The probability of obtaining a ratio this large or larger by chance alone is 0.0318. That is, if the two samples came from populations with equal variance, there is a small probability (.0318) of obtaining a ratio of our sample variances of 12.40 or larger by chance. We should therefore use the t-value appropriate for groups with unequal variance. The rule of thumb here is to use the t-value, df, and probability labeled **unequal** if the probability from the F-test is **less** then .05.

## B. RANDOM ASSIGNMENT OF SUBJECTS

In our discussion of t-tests, we said that we randomly assigned subjects to either a treatment or control group. This is actually a very important step in our experiment and we can use SAS software to provide us with a method for doing this.

We could take our volunteers one by one, flip a coin, and decide to place all the "heads" in our treatment group and the "tails" in our control group. This is acceptable but we would prefer a method that ensures an equal number of subjects in each group. One method would be to place all the subjects' names in a hat, mix them up, and pull out half the names for treatment subjects and the others for controls. This is essentially what we will do in our SAS program. The key to the program is the SAS random number function.

The function UNIFORM(0) will generate a pseudo-random number in the interval from 0 to 1. The zero argument of the UNIFORM function is called the **seed**. This is a number that is used the first time the function is called, to start the random number series. A zero argument instructs the SAS program to use a value from the internal **time clock** as the seed. Using the time clock as a seed will generate a different series of random numbers each time you run the program. You may want to supply your own seed instead. It should be a five-, six-, or seven-digit odd number. Each time you run the program with your own seed, you will generate the **same series of random numbers**. This is sometimes desirable, but as a rule, use a zero seed unless you have a specific need for a repeatable series of random numbers. We will create a SAS data set with our subjects' names (or we could use subject numbers instead) and assign a random number to each subject. We can then split the group in half (or in any number of subgroups) using a special feature of PROC RANK. Here is the program:

```
PROC FORMAT;
    VALUE GRPFMT 0=CONTROL 1=TREATMENT;
DATA RANDOM;
INPUT SUBJ NAME $20.;
GROUP=UNIFORM(0);
CARDS;
1 CODY
2 SMITH
3 HELM
4 BERNHOLC
  etc.
PROC RANK GROUPS=2;
    VAR GROUP;
PROC SORT;
    BY NAME;
PROC PRINT;
    TITLE 'SUBJECT GROUP ASSIGNMENTS';
    ID NAME;
    VAR SUBJ GROUP;
    FORMAT GROUP GRPFMT.;
```

The key to this program is the UNIFORM function, which assigns a random number from 0 to 1 to each subject. The GROUPS=2 option of PROC RANK divides the subjects into two groups (0 and 1), depending on the value of the random variable GROUP. Values below the median become 0; those above the median become 1. The GRPFMT format assigns the labels "CONTROL" and "TREATMENT" using values of 0 and 1, respectively. We can use this program to assign our subjects to any number of groups by changing the "GROUPS=" option of PROC RANK to indicate the desired number of groups. Sample output from this program is shown below:

```
SUBJECT GROUP ASSIGNMENTS

  NAME    SUBJ GROUP

BERNHOLC   4   CONTROL
CODY       1   CONTROL
HELM       3   TREATMENT
SMITH      2   TREATMENT
```

Although this may seem like a lot of work just to make random assignments of subjects, we recommend that this or an equivalent procedure be used for assigning subjects to groups. Other methods such as assigning every other person to the treatment group can result in unsuspected bias.


## C. TWO INDEPENDENT SAMPLES: DISTRIBUTION FREE TESTS

There are times when the assumptions for using a t-test are not met. One common problem is that the data are not normally distributed. For example, suppose we collected the following numbers in a psychology experiment that measured the response to a stimulus:

0 6 0 5 7 6 9 4 8 0 7 0 5 6 6 0 0

A frequency distribution would look like this:

```
        x
        x
        x       x
        x       x
        x      xxx
        x      xxxxxx
        --------------------
        0123456789
```

What we are seeing is probably due to a threshold effect. The response is either zero (the stimulus is not detected) or, once the stimulus is detected, the average response is about 6. Data of this sort would artificially inflate the standard deviation (and thus the standard error) of the sample and make the t-test more conservative. However, we would be safer to use a non-parametric test (a test that does not assume a normal distribution of data).

Another common problem is that the data values may only represent ordered categories. Scales like 1=very mild, 2=mild, 3=moderate, 4=strong, 5=severe, et reflect the strength of a response, but we cannot say that a score of 4 (strong) is worth twice the score of 2 (mild). Scales like these are referred to as **ordinal scales**. (Most of the scales we have been using up till now have been **interval** or **ratio scales**.) We will need a nonparametric test to analyze differences in central tendencies for ordinal data. Finally, for very small samples, nonparametric tests are often more appropriate since assumptions concerning distributions are hard to determine.

SAS software provides us with several nonparametric two-sample tests. Among these are the Wilcoxon rank-sum test (equivalent to the Mann-Whitney U-test) and the median test for two samples.

Consider the following experiment. We have two groups, A and B. Group B has been treated with a drug to prevent tumor formation. Both groups are exposed to a chemical that will encourage tumor growth. The masses (in grams) of tumors in groups A and B are

```
A: 3.1 2.2 1.7 2.7 2.5
B: 0.0 0.0 1.0 2.3
```

Are there any differences in tumor mass between groups A and B? We will choose a nonparametric test for this experiment because of the absence of a normal distribution and the small sample sizes involved. The Wilcoxon test first puts all the data (groups A and B) in increasing order (with special provisions for ties), retaining the group identity. In our experiment we have

```
MASS    0.0 0.0 1.0 1.7 2.2 2.3 2.5 2.7 3.1
GROUP    B   B   B   A   A   B   A   A   A
RANK    1.5 1.5  3   4   5   6   7   8   9
```

The sums of ranks for the A's and B's are then computed. We have

```
SUM RANKS A = 4+5+7+8+9 = 33
SUM RANKS B = 1.5+1.5+3+6 = 12
```

If there were smaller tumors in group B, we would expect the B's to be at the lower end of the rank ordering and therefore have a smaller sum of ranks than the A's. Is the sum of ranks for group A sufficiently larger than the sum of ranks for group B so that the probability of the difference occurring by chance alone is small (less than .05)? The Wilcoxon test gives us the probability that the difference in rank sums that we obtained occurred by chance.

For the median test, we first calculate the median of groups A and B combined. In our example, the middle score is 2.2. We then construct a table that shows how many measurements in groups A and B fall below or above the median:

```
                        A    B
                     ----------
    BELOW MEDIAN     |  1 |  4 |
                     ----------
    MEDIAN OR ABOVE  |  3 |  1 |
                     ----------
```

A chi-square statistic and its associated probability is then calculated. The SAS procedure for computing the probability for the median test is valid for small sample sizes, so the example above will be computed correctly.

In general, the Wilcoxon test will be more powerful (give a smaller p-value) than the median test. The reason is that the median test only places values in one of two categories (below or above the median) while the

Wilcoxon test is concerned with the order of all the data values. For even moderate sample sizes, the Wilcoxon test is almost as powerful as its parametric equivalent, the t-test. Thus, if there is a question concerning distributions or if the data are really ordinal, you should not hesitate to use the Wilcoxon test instead of the t-test.

The program to analyze this experiment using both the Wilcoxon and median tests is shown below:

```
DATA;
INPUT GROUP $ MASS @@;
CARDS;
A 3.1 A 2.2 A 1.7 A 2.7 A 2.5
B 0.0 B 0.0 B 1.0 B 2.3
PROC NPAR1WAY WILCOXON MEDIAN;
   CLASS GROUP;
   VAR MASS;
```

First, we have introduced a new feature on the INPUT statement. Normally, when a SAS program has finished reading an observation, it goes to a new line of data for the next observation. The two '@' signs at the end of the INPUT statement instruct the program to **"hold the line"** and not automatically go to a new line of data for the next observation. This way, we can put data from several observations on one line. By the way, a single @ sign will hold the line for another INPUT statement but will go to the next line when the CARDS statement is encountered.

PROC NPAR1WAY performs the nonparametric tests. The options WILCOXON and MEDIAN request these particular tests. The CLASS and VAR statements are identical to the CLASS and VAR statements of the t-test procedure.

The output from the NPAR1WAY procedure follows:

```
ANALYSIS FOR VARIABLE MASS CLASSIFIED BY VARIABLE GROUP

MIDRANKS WERE USED FOR TIES
WILCOXON SCORES (RANK SUMS)
                            SUM OF    EXPECTED    STD DEV       MEAN
LEVEL                N      SCORES    UNDER H0    UNDER H0      SCORE

A                    5      33.00      25.00        4.08         6.60
B                    4      12.00      20.00        4.08         3.00
WILCOXON 2-SAMPLE TEST (NORMAL APPROXIMATION)
S=   12.00      Z=-1.8371      PROB >|Z|=0.0662
T-TEST APPROX. SIGNIFICANCE=0.1035
KRUSKAL-WALLIS TEST (CHI-SQUARE APPROXIMATION)
CHISQ=  3.84    DF=  1    PROB > CHISQ=0.0500
MEDIAN SCORES (NUMBER POINTS ABOVE MEDIAN)
                            SUM OF    EXPECTED    STD DEV       MEAN
LEVEL                N      SCORES    UNDER H0    UNDER H0      SCORE

A                    5       3.00       2.22        0.79         0.60
B                    4       1.00       1.78        0.79         0.25
MEDIAN 2-SAMPLE TEST (NORMAL APPROXIMATION)
S=    1.00      Z=-0.9838      PROB >|Z|=0.3252
MEDIAN 1-WAY ANALYSIS (CHI-SQUARE APPROXIMATION)
CHISQ=0.98          DF=  1     PROB > CHISQ=0.3222
```

The  sum of ranks for groups A and B are shown,  as well as their expected values.  The probability that the medians differ by chance alone is .0662.

The median test results are shown next.  The number of subjects in groups A and B above the median are shown first,  along with an expected value.  Two probabilities are computed:  The first (.3252) resulting from a normal approximation  method  and the second (.3222)  from  the chi-square  statistic.   The two values should  be  very close.   You may decide to use the larger p-value  if  a conservative test is desired.

Notice  how much smaller the p-value from the  Wilcoxon  test  is  compared  to  the  median  test  value. However,  if our alpha level is .05, then even using the Wilcoxon test we cannot reject the null hypothesis  that the antitumor drug is effective.

## D. ONE-TAILED VERSUS TWO-TAILED TESTS

When we conduct an experiment like the tumor example in Section B, we have a choice in the way we state the alternate hypothesis. In our example, the null hypothesis was that the mass of tumors was the same for groups A and B. The alternate hypothesis was that groups A and B were not the same. We would reject the null hypothesis if A>B or B>A. This type of hypothesis requires a two-tailed test. If we were using a t-test, we would have to consider absolute values of t greater than the critical value. If our alpha level is .05 then we have .025 from each **tail** of the t distribution.

In many research studies, the researcher has a reasonable expectation of the results. If we are testing a new drug against a placebo, we **expect** the drug to be better. If we are testing whether people prefer blue widgets to red widgets, then we probably do **not** have an expectation either way. Whenever a directional alternative hypothesis (e.g., B > A) can be justified from the **substantive issues** in the study, then a one-tailed test can be used. With a one-tailed test, the 5% of the curve associated with the .05 alpha level can all be located in one tail, which increases the power of the study (i.e., makes it more likely to find a significant difference when, in fact, one exists.) If our tumor example had been stated as a one-tailed test, we could have divided the p-value by 2, giving p=.0331 for the Wilcoxon test probability. **The decision to do a one-tailed test should be based on an understanding of the experimental procedure and not as a method of reducing the p-value below the .05 level.**

## E. PAIRED T-TESTS (RELATED SAMPLES)

Our t-test example in section A had subjects randomly assigned to a control or treatment group. Therefore, the groups could be considered to be independent.

There are many experimental situations where each subject receives both treatments. For example, each subject in the earlier example could have been measured in the absence of the drug (control value) and after having received the drug (treatment value). The response times for the control and treatment groups would no longer be independent. We would expect a person with a very short response time in the control situation to also have a short response time after taking the drug (compared to the other people who took the drug). We would expect a positive correlation between the control and treatment values.

Our regular t-test cannot be used here since the groups are no longer independent. A variety of the t-test referred to as a paired t-test is used instead. The differences between the treatment and control times are computed for each subject. If most of the differences are positive, we suspect that the drug lengthens reaction time. If most are about zero, the drug has no effect. The paired t-test essentially computes a mean and standard error of the **differences** and determines the probability that the absolute value of the mean difference was greater than zero by chance alone.

Before we program this problem, it should be mentioned that this would be a very poor experiment. What if the response time increased because the subjects became tired? If each subject were measured twice, without being given a drug, would the second value be different because of factors other than the treatment? One way to control for this problem is to take half of the subjects and measure their drug response time first and the control value later, after the drug has worn off. We will see a better way to devise experiments to handle this problem in Chapter 7 (Repeated Measures Designs).

Experiments that measure the same subject under different conditions are sometimes called repeated measures experiments. They do have the problem just stated: one measurement might affect the next. However, if this can be controlled, it is much easier to show treatment effects with smaller samples compared to a regular t-test.

If we took two independent groups of people, we would find that within the control group or the treatment group there would be variation in response times because of individual differences. However, if we measure the same subject under two conditions, even if that person has much longer or shorter response times than the other subjects, the DIFFERENCE between the scores should approximate the difference for other subjects. Thus, each subject acts as his own control and therefore controls some of the natural variation between subjects.

The SAS system does not include a paired t-test as part of PROC TTEST. We will need to compute the difference scores ourselves and then use PROC MEANS to compute the probability that the difference is significantly different from zero.

Our data are arranged like this:

| SUBJECT | CONTROL VALUE | TREATMENT VALUE |
|---------|---------------|-----------------|
| 1 | 90 | 95 |
| 2 | 87 | 92 |
| 3 | 100 | 104 |
| 4 | 80 | 89 |
| 5 | 95 | 101 |
| 6 | 90 | 105 |

Note: Subjects 1-3 were measured in control/treatment order, subjects 4-6 were measured in treatment/control order.

The program is written as follows:

```
DATA;
INPUT CTIME TTIME;
DIFF=TTIME-CTIME;
CARDS;
90 95
87 92
100 104
80 89
95 101
90 105
PROC MEANS N MEAN STDERR T PRT;
   VAR DIFF;
```

The variable names CTIME and TTIME were chosen to represent the response times in the control and treatment conditions, respectively. For each observation, we are calculating the difference between the treatment and control times, by creating a new variable called DIFF.

PROC MEANS is followed by a list of options. N, MEAN, and STDERR cause the number of subjects, the mean, and the standard error of the mean to be printed. In our case, these statistics will be computed for the variable DIFF. The options T and PRT will give us the t-value and its associated probability, testing if the variable (DIFF) is significantly different from zero. The output is shown below:

| VARIABLE | N | MEAN | STD ERROR OF MEAN | T | PR>|T| |
|----------|---|------|-------------------|---|--------|
| DIFF | 6 | 7.33333333 | 1.68654809 | 4.35 | 0.0074 |

In this example, the mean difference is positive (7.3) and the probability of the difference occurring by chance is .0074. We can state that response times are longer under the drug treatment compared to the control values. Had the mean difference been negative, we would state that response times were shorter under the drug treatment, because DIFF was computed as treatment time minus control time.

# Chapter 6 Hypothesis Testing (More than Two Groups)

## A. ONE-WAY ANALYSIS OF VARIANCE

We have analyzed experiments with two groups (control and treatment) using a t-test. Now, what if we have more than two groups? Take the situation where we have three treatment groups: A, B, and C. It was once the practice to use t-tests with this design, comparing A with B, A with C, and B with C. With four groups (ABCD) we would have comparisons AB, AC, AD, BC, BD, and CD. As the number of groups increases, the number of possible comparisons grows rapidly. What is wrong with this procedure of multiple comparisons? Any time we show the means of two groups to be significantly different, we state a probability that the difference occurs by chance alone. Suppose we make 20 multiple comparisons and each comparison is made at the .05 level. Even if there are no **REAL** differences between the groups, we expect to find one comparison significant at the .05 level. (The actual probability of at least one significant difference by chance alone is .64.) The more comparisons that are made, the greater the likelihood of finding a pair of means significantly different by chance alone.

The method used today for comparisons of three or more groups is called **An**alysis of **Va**riance (**ANOVA**). This method has the advantage of testing whether there are **any** differences between the groups with a single probability associated with the test. The hypothesis tested is that **all** groups have the same mean. Before we present an example, we should note that there are several **assumptions** that should be met before an analysis of variance is used. Essentially, the same assumptions for a t-test need to be met when conducting an ANOVA. That is, we must have **independence** between groups (unless a **repeated measures design** is used), the sampling distributions of sample means must be **normally distributed**, and the groups should have nearly equal variances (called **homogeneity of variance**). The analysis of variance technique is said to be **robust**. This is a term used by statisticians which means that the assumptions can be violated somewhat and the technique can still be used. So, if the distributions are not perfectly normal or if the variances are unequal, we may still  use analysis of variance.  The point at which the assumptions  are violated beyond  the  point where ANOVA can  be used is somewhat subjective,  and,  if in doubt, see  your local statistician.  Winer (see Chapter 1  for the reference) has an excellent discussion of the effect of homogeneity of  variance  violations  and  the use of analysis of variance.  Balanced designs  (those with the same number of subjects under  each  of the experimental conditions)  are  preferred  to  unbalanced  designs, especially when the group variances are unequal.

Consider the following experiment:

We randomly assign 15  subjects to three  treatment groups X,Y,  and Z (with 5 subjects per treatment). Each of the three groups has received a  different  method of speed reading instruction.  A reading test is given and the number of words per minute is recorded for each sub-ject. The following data are collected:

|   X   |   Y   |   Z   |
|-------|-------|-------|
| 700   | 480   | 500   |
| 850   | 460   | 550   |
| 820   | 500   | 480   |
| 640   | 570   | 600   |
| 920   | 580   | 610   |

The null hypothesis is that mean(X)  =  mean(Y)  = mean(Z).  The alternative hypothesis is that the  means are not all equal.  The means of groups X,Y,  and Z  are 786,  518,  and 548, respectively. How do we know if the means  obtained are different because of differences  in the  reading  programs  or because  of  random  sampling error? By chance,  the five subjects we choose for group X might be faster readers than those chosen for groups Y and Z.

In  our example,  the mean reading speed of all  15 subjects (called the GRAND MEAN)  is 617.33. Now we nor-mally think of a subject's score as whatever it  happens to be,  580  is 580.  But we could also think of 580  as being 37.33 points lower than the grand mean (580-617.33 = -37.33).

We might now ask the question, "what causes scores to vary from the grand mean?" In this example, there are two possible sources of variation. The first source of variation is the training method (X,Y, or Z). If X is a far superior method, then we would expect subjects in X to have higher scores, in general, than subjects in Y or Z. When we say "higher scores in general" we mean something quite specific. We mean that being a member of group X causes one's score to increase by so many points.

The second source of variation is due to the fact that individuals are different. Therefore, within each group there will be variation. We can think of a formula to represent each person's score:

THE PERSON'S = THE GRAND   AN ADDITION OR           AN ADDITION OR
   SCORE         MEAN   +  SUBTRACTION FROM     +   SUBTRACTION DE-
                           THE GRAND MEAN DE-       PENDING ON THE
                           PENDING ON WHICH         INDIVIDUAL'S
                           GROUP THE PERSON         VARIABILITY.
                           IS IN.

Now that we have the ideas down, let's return briefly to the mathematics.

It turns out that the mathematics are simplified if, instead of looking at differences in scores from the grand mean, we look instead at the square of the differences. The sum of all the squared deviations is called the total SUM OF SQUARES or **SS total**.

To be sure this is clear, we will calculate the total SS in our example. Subtracting the grand mean (617.33) from each score, squaring the differences (usually called **deviations**), and adding up all the results, we have

$$\text{SS total} = (700-617.33)^2 + (850-617.33)^2 + \ldots + (610-617.33)^2$$

As we mentioned earlier, we can separate the total variation into two parts: one due to differences in the reading methods (often called SUM OF SQUARES BETWEEN (groups)) and the other due to the normal variations between subjects (often called the SUM OF SQUARES ERROR). Note that the word ERROR here is not the same as "mistake." It simply means that there is variation in the scores that we cannot attribute to a specific variable. Some statisticians call this **residual** instead of **error**.

Our intuition tells us that if there is a large variation between the group means compared to variation within each group, then the means could be considered to be different because of differences in the reading methods. If we take the "average" sum of squares due to group differences (MEAN SQUARE group) divided by the "average" sum of squares due to subject differences (MEAN SQUARE error), the result is called an F ratio.

$$F = \frac{MS\ group}{MS\ error}$$

If the variation **BETWEEN** the groups is large compared to the variation **WITHIN** the groups, this ratio will be large. If the null hypothesis is true, the F statistic will be equal to 1.00 on the average. Just how far away from 1.00 is **too** far away to be attributable to chance is a function of the number of groups and the number of subjects in each group. SAS analysis of variance procedures will give us the F ratio and the probability of obtaining a value of F this large or larger by chance alone.

The box below contains a more detailed explanation of how an F ratio is computed. You may wish to skip the box for now.

Consider a one-way ANOVA with three groups (A, B, C) and three subjects within each group. The design is shown below:

```
                A       B       C
                50      70      20
  Within    |   40      80      15
  Group     |   60      90      25

  MEANS         50      80      20      50  GRAND MEAN
                <-------------->
                   Between
                    Groups
```

If we want to estimate the within-group variance (also called ERROR variance), we take the deviation of each score from the group mean, square the result, and add up the squared deviations for each of the groups and divide the result by the degrees of freedom.  (The number of degrees of freedom is N-k, where N is the total number of subjects and k is the number of groups.) In the example above, the within group variance is equal to

$$\overbrace{[(50\text{--}50)^2 \; + \; (40\text{--}50)^2 \; + \; (60\text{--}50)^2}^{A} \; + \; (70\text{--}80)^2 \; + \ldots$$

$$+ \; (25\text{--}20)^2 \; ]/6 = 450/6 = 75.0$$

This within-group variance estimate (75.0) can be compared to the between-group variance. The between-group variance is obtained by taking the squared deviations of each group mean from the grand mean, multiplying each deviation by the number of subjects in a group, and dividing by the degrees of freedom (k-1). In our example the between-group variance is

$$[3(50\text{--}50)^2 + 3(80\text{--}50)^2 + 3(20\text{--}50)^2]/2 = 2700$$

If the null hypothesis is true, the between-group variance estimate will be close to the within-group variance and the ratio

$$F = \frac{\text{BETWEEN GROUP VARIANCE}}{\text{WITHIN GROUP VARIANCE}}$$

will be close to 1. In our example, F = 2700/75 = 36.0 with probability of .0005 of obtaining a ratio this large or larger by chance alone.

We can write the following program (using our reading speed data):

```
DATA;
INPUT GROUP $ WORDS;
CARDS;
X 700
X 850
  .
  .
  .
Z 610
PROC PRINT;
PROC ANOVA;
    TITLE 'ANALYSIS OF READING DATA';
    CLASSES GROUP;
    MEANS GROUP;
    MODEL WORDS=GROUP;
```

The variable name following the word **CLASSES** defines the **independent** variable. In our case, the variable GROUP will have values of X,Y, or Z. The MEANS statement is optional (but in most instances you will want it). In this example, MEANS GROUP will give us the mean value of the dependent variable (WORDS) for each level of group. Finally, following the word **MODEL** is our **dependent** variable and, to the right of the equals sign, the **independent** variable. Output from this program is shown below:

```
ANALYSIS OF READING DATA

ANALYSIS OF VARIANCE PROCEDURE

CLASS LEVEL INFORMATION

CLASS      LEVELS     VALUES

GROUP         3       X Y Z


NUMBER OF OBSERVATIONS IN DATA SET = 15
```

```
DEPENDENT VARIABLE: WORDS

SOURCE                    DF        SUM OF SQUARES        MEAN SQUARE

MODEL                      2       215613.33333333     107806.66666667

ERROR                     12        77080.00000000       6423.33333333

CORRECTED TOTAL           14       292693.33333333


MODEL F =              16.78                          PR > F = 0.0003


R-SQUARE               C.V.             STD DEV          WORDS MEAN

0.736653              12.9826          80.14570065       617.33333333


SOURCE                    DF             ANOVA SS    F VALUE    PR > F

GROUP                      2       215613.33333333     16.78    0.0003


MEANS

GROUP          N        WORDS

X              5      786.000000
Y              5      518.000000
Z              5      548.000000
```

The analysis of variance table shows us the sources of variation, the SUM OF SQUARES and MEAN SQUARE due to this source of variation, and the F ratio. The first source of variation is **model**. Model represents the totality of **all** the independent variables (often called "factors" in ANOVA) which we designed into the study. In this example, since we have only one independent variable (GROUP), the term MODEL will be synonymous with GROUP. We see that our F value is 16.78 and the probability of obtaining a value this large by chance is 0.0003. We would therefore reject the null hypothesis and conclude that the reading instruction methods were not all equivalent.

Now that we know the reading methods are different, we want to know what the differences are. Is X better than Y or Z? Are the means of groups Y and Z so close that we cannot consider them different? In general, methods used to find group differences after the null hypothesis has been rejected are called **post hoc** or **a posteriori** tests. SAS software provides us with a variety of these tests to investigate differences between levels of our independent variable. These include **Duncan's multiple-range test**, the **Student-Newman-Keuls' multiple range test**, **least significant difference test**, **Tukey's studentized range test**, **Scheffe's multiple-comparison procedure**, and others. To request a post hoc test, place the SAS option name for the test you want, following a slash (/) on the MEANS statement. The SAS names for the post hoc tests previously listed are **DUNCAN, SNK, LSD, TUKEY,** and **SCHEFFE**, respectively. In practice, it is easier to include the request for a post hoc test at the same time we request the analysis of variance. If the analysis of variance is not significant, **WE SHOULD NOT LOOK FURTHER AT THE POST HOC TEST RESULTS.** Our examples will use Duncan's multiple-range test for post hoc comparisons. You may use any of the available methods in the same manner. Winer (see Chapter 1) is an excellent reference for analysis of variance and experimental design. A discussion of most of these post hoc tests can be found there.

For our example we have


▌    MEANS GROUP / DUNCAN;


Unless we specify otherwise, the differences between groups are evaluated at the .05 level. Alpha levels of .1 or .01 may be specified by following the post hoc option name with ALPHA=.1 or ALPHA=.01. For example, to specify an alpha level of .1 for a Scheffe test, we would have

MEANS GROUP / SCHEFFE ALPHA=.1;                                    ▌

Here is the output from the Duncan procedure in our example:

DUNCAN'S MULTIPLE RANGE TEST FOR VARIABLE WORDS

MEANS WITH THE SAME LETTER ARE NOT SIGNIFICANTLY DIFFERENT.

ALPHA LEVEL=.05          DF=12          MS=6423.33

| GROUPING | MEAN | N | GROUP |
|----------|------|---|-------|
| A | 786.000000 | 5 | X |
| B | 548.000000 | 5 | Z |
| B |  |  |  |
| B | 518.000000 | 5 | Y |

The Duncan procedure uses the following method to show group differences:

On the right are the group identifications. The order is determined by the group means, from highest to lowest. At the far left is a column labeled "GROUPING." Any groups that are not significantly different from one another will have the same letter in the GROUPING column. In our example, the Y and Z groups both have the letter 'B' in the GROUPING column and are therefore not significantly different. The letter 'B' between GROUP Z and GROUP Y is there for visual effect. It helps us realize that groups Y and Z are not significantly different (at the .05 level). Group X has an A in the grouping column and is therefore significantly different (p < .05) from the Y and Z groups.

From this Duncan's test we conclude that

1. Method X is superior to both methods Y and Z.
2. Methods Y and Z are not significantly different.

How would we describe the statistics used and the results of this experiment in a journal article? Although there is no "standard" format, we will suggest one approach here. The key is clarity. Here is our suggestion:

METHOD

We compared three reading methods: (1) Smith's Speed Reading Course, (2) Standard Method, and (3) Evelyn Tree's Institute. Fifteen subjects were randomly assigned to one of the three methods. At the conclusion of training, a standard reading test (Cody Count the Words Test version 2.1) was administered.

RESULTS

The mean reading speed for the three methods was

| METHOD | READING SPEED (Words per minute) |
|---|---|
| (1) Smith's | 786 |
| (2) Standard | 518 |
| (3) Tree's | 548 |

A one-way analysis of variance was performed. The F-value was 16.78 (df=2,12, p=.0003). A Duncan multiple range test (p=.05) showed that Smith's method was significantly superior to either Tree's or the Standard method. Tree's and the Standard method were not significantly different from each other.

The results of the Duncan multiple range test can easily be described in words when there are only three groups. With four or more groups, especially if the results are complicated, we can use another method. Consider the following results of a Duncan test on an experiment with 4 treatment groups:

| GROUPING | MEAN | N | GROUP |
|----------|------|---|-------|
| A        | 80   | 10 | 1 |
| A        |      |    |   |
| A  B     | 75   | 10 | 3 |
|    B     |      |    |   |
|    B     | 72   | 10 | 2 |
|          |      |    |   |
| C        | 60   | 10 | 4 |

We see that groups 1 and 3 are not significantly different. (They both have the letter "A" in the GROUPING column). Neither are groups 2 and 3. But 1 and 2 are! Remember that **"not significantly different"** does not mean **"equal."** Finally, group 4 is significantly different from all the other groups ($p < .05$). We can describe these results in a journal article like this:

GROUP

| | 1 | 3 | 2 | 4 |
|------|----|----|----|----|
| MEAN | 80 | 75 | 72 | 60 |

DUNCAN MULTIPLE RANGE TEST EXAMPLE

**Any two groups with a common underscore are not significantly different ($p < .05$).**

## B. ANALYSIS OF VARIANCE: TWO INDEPENDENT VARIABLES

Suppose we ran the same experiment comparing reading methods, using 15 male subjects and 15 female subjects. In addition to comparing reading instruction methods, we could compare male vs. female reading speeds. Finally, we might want to see if the effects of the reading methods are the same for males and females.

This experimental design is called a two-way analysis of variance. The "two" refers to the fact that we have two independent variables: GROUP and SEX. We can picture this experiment as follows:

```
                          GROUP

              X            Y            Z
         -----------------------------------
         |                             |
         |  700   |   480   |   500   |
         |  850   |   460   |   550   |
  MALE   |  820   |   500   |   480   |
         |  640   |   570   |   600   |
         |  920   |   580   |   610   |
SEX   ----------------------------------|
         |  900   |   590   |   520   |
         |  880   |   540   |   660   |
 FEMALE  |  899   |   560   |   525   |
         |  780   |   570   |   610   |
         |  899   |   555   |   645   |
         -----------------------------------
```

In this design we have each of the three reading instruction methods for each level of SEX (male/female). Designs of this sort are called **factorial** designs. The combination of GROUP and SEX is called a cell. For this example, males in group X constitute a cell. In general, the number of cells in a factorial design would be the number of levels of one independent variable times the number of levels of the other independent variable. Three levels of GROUP times two levels of SEX = 6 cells in our case.

The total sum of squares is now divided or partitioned into four components. We have the sum of squares due to GROUP differences and the sum of squares due to SEX differences. The combination of GROUP and SEX provides us with another source of variation (called an interaction), and finally, the remaining sum of squares is attributed to error. We will discuss the interaction term later in this chapter.

Since there are the same number of subjects in each cell, the design is said to be "balanced" (some statisticians call this "orthogonal"). When we have **more than one independent variable** in our analysis of variance, we can use PROC ANOVA **only for balanced designs.** For unbalanced designs, PROC GLM (general linear model) is used instead. The programming of our balanced design experiment is similar to the one-way analysis of variance. Here is the program:

```
DATA;
INPUT GROUP $ SEX $ WORDS;
CARDS;
X M 700
  .
  .
Z F 645
PROC ANOVA;
   TITLE 'ANALYSIS OF READING DATA';
   CLASSES GROUP SEX;
   MEANS GROUP|SEX / DUNCAN;
   MODEL WORDS = GROUP|SEX;
```

As before, following the word CLASSES is a list of independent variables. The vertical line between GROUP and SEX in the MEANS and MODEL statements indicates that we have a factorial design (also called a crossed design). Some computer terminals may not have the "|" symbol on the keyboard. In this case, the term GROUP|SEX can be written as

GROUP SEX GROUP*SEX

The "|" symbol is especially useful when we have higher-order factorial designs such as GROUP|SEX|DOSE. Written out the long way, this would be

GROUP SEX DOSE GROUP*SEX GROUP*DOSE
SEX*DOSE GROUP*SEX*DOSE

That is, each variable, and every two- and three-way interaction term has to be specified.

Let's study the output of the previous example carefully to see what conclusions we can draw about our experiment. The first portion of the output is shown below:

```
ANALYSIS OF READING DATA

ANALYSIS OF VARIANCE PROCEDURE

CLASS LEVEL INFORMATION

CLASS      LEVELS    VALUES

GROUP        3       X Y Z

SEX          2       F M


NUMBER OF OBSERVATIONS IN DATA SET = 30

DEPENDENT VARIABLE: WORDS

SOURCE                    DF       SUM OF SQUARES            MEAN SQUARE

MODEL                      5       531436.16666667      106287.23333333

ERROR                     24       106659.20000000        4444.13333333

CORRECTED TOTAL           29       638095.36666667


MODEL F =              23.92                        PR > F = 0.0001


R-SQUARE              C.V.            STD DEV              WORDS MEAN

0.832848             10.3126        66.66433329          646.43333333


SOURCE                    DF            ANOVA SS     F VALUE     PR > F

GROUP                      2       503215.26666667     56.62     0.0001
SEX                        1        25404.30000000      5.72     0.0250
GROUP*SEX                  2         2816.60000000      0.32     0.7314
```

The  top portion labeled "CLASS LEVEL  INFORMATION"
indicated  our two independent variables and the  levels
of each. The analysis of variance table shows us the SUM
OF SQUARES and MEAN SQUARE for the entire model and  the
error.  This overall F value (23.92) and the probability
p=.0001  shows us how well the model (as a whole)   ex-
plains the variation about the grand mean. This could be

very important in certain types of studies where we want to create a general predictive model. In this case, we are more interested in the detailed sources of variation (GROUP, SEX, and GROUP*SEX).

Each source of variation in the table has an F value and the probability of obtaining a value of F this large or larger by chance. In our example, the GROUP variable was significant at .0001 and SEX at .0250. Since there are only two levels of SEX, we do not need the Duncan test to claim that males and females are significantly different with respect to reading speed (p=.025).

In a two-way analysis of variance, when we look at GROUP effects, we are comparing GROUP levels without regard to SEX. That is, when the GROUPs are compared, we combine the data over males and females. Conversely, when we compare males to females, we combine data from the three treatment groups.

The term GROUP*SEX is called an **interaction** term. If group differences were **not** the same for males and females, we would have a significant interaction. For example, if males did better with method A compared to method B while females did better with B compared to A, we would expect a significant interaction. In our example, the interaction between GROUP and SEX was not significant (p=.73). (Our next example will show a case where there is a significant interaction.)

The portion of the output resulting from the "MEANS GROUP|SEX / DUNCAN" request is shown next:

```
DUNCAN'S MULTIPLE RANGE TEST FOR VARIABLE WORDS

MEANS WITH THE SAME LETTER ARE NOT SIGNIFICANTLY DIFFERENT.

ALPHA LEVEL=.05        DF=24        MS=4444.13


        GROUPING                MEAN      N    GROUP

           A                 828.800000   10    X

           B                 570.000000   10    Z
           B
           B                 540.500000   10    Y

DUNCAN'S MULTIPLE RANGE TEST FOR VARIABLE WORDS

MEANS WITH THE SAME LETTER ARE NOT SIGNIFICANTLY DIFFERENT.

ALPHA LEVEL=.05        DF=24        MS=4444.13


        GROUPING                MEAN      N    SEX

           A                 675.533333   15    F

           B                 617.333333   15    M

MEANS

GROUP    SEX          N       WORDS

  X       F           5     871.600000
  X       M           5     786.000000
  Y       F           5     563.000000
  Y       M           5     518.000000
  Z       F           5     592.000000
  Z       M           5     548.000000
```

The first comparison shows group X significantly different ($p < .05$) from Y and Z. The second table shows that females have significantly higher reading speeds than males. We already know this because SEX was a significant main effect ($p=.025$) and there are only two levels of SEX. Following the two Duncan tests are the mean reading speeds for each combination of GROUP and

SEX.   These values are the means of the 6 cells in   our experimental design.


## C. INTERPRETING SIGNIFICANT INTERACTIONS

We  will  now discuss an example that  has  a  sig- nificant interaction term.  We have two groups of  chil- dren.  One group is considered normal,  the other,  hy- peractive.  Each group of children is randomly  divided, with one half receiving a placebo and the other  a  drug called ritalin.  A measure of activity is determined for each of the children. The following data are collected:

```
                    PLACEBO      RITALIN
                 ---------------------------
                      50           67
                      45           60
     NORMAL           55           58
                      52           65
     --------------------------------------------
                      70           51
     HYPERACTIVE      72           57
                      68           48
                      75           55
```

We  will  name the variables in  this  study  GROUP (NORM or HYPER), DRUG (PLACEBO or RITALIN), and ACTIVITY (activity  score).  Since the design is balanced  (same number of subjects per cell), we can use PROC ANOVA. The PROC statements are written like this:

```
PROC ANOVA;
     TITLE 'ACTIVITY STUDY';
     CLASSES GROUP DRUG;
     MEANS GROUP|DRUG;
     MODEL ACTIVITY = GROUP|DRUG;
```

This  is  a two-way analysis of variance  factorial design just like the last example.  Again,  the vertical bar between GROUP and DRUG in the MEANS and MODEL state- ments indicates that we have a factorial design and that

GROUP and DRUG are crossed. Notice that we do not need to request a Duncan test since there are only two levels of each independent variable.

A portion of the output is shown below:

| SOURCE | DF | ANOVA SS | F VALUE | PR > F |
|---|---|---|---|---|
| GROUP | 1 | 121.00000000 | 8.00 | 0.0152 |
| DRUG | 1 | 42.25000000 | 2.79 | 0.1205 |
| GROUP*DRUG | 1 | 930.25000000 | 61.50 | 0.0001 |

The first thing to notice is that there is a strong GROUP*DRUG interaction term (p=.0001). When this occurs, we must be careful about interpreting any of the main effects (GROUP and DRUG in our example). That is, we must first understand the nature of the interactions before we look at main effects.

By looking more closely at the interaction between GROUP and DRUG, we will see why the main effects shown in the analysis of variance table can be misleading. The best way of explaining a two-way interaction is to take the cell means and plot them. These means can be found in the portion of the output from the MEANS request. The portion of the output containing the cell means is shown below:

| Level of GROUP | Level of DRUG | N | Mean | SD |
|---|---|---|---|---|
| | | | ------------ACTIVITY------------ | |
| HYPER | PLACEBO | 4 | 71.2500000 | 2.98607881 |
| HYPER | RITALIN | 4 | 52.7500000 | 4.03112887 |
| NORM | PLACEBO | 4 | 50.5000000 | 4.20317340 |
| NORM | RITALIN | 4 | 62.5000000 | 4.20317340 |

We can use this set of means to plot an interaction graph. We pick one of the independent variables (we will choose DRUG) to go on the x-axis and then plot means for each level of the other independent variable (GROUP). The resulting graph is shown below:

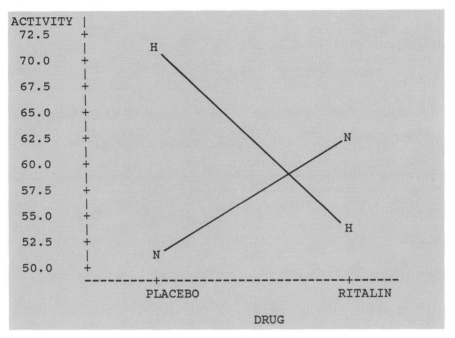

```
ACTIVITY |
   72.5   +
         |         H
   70.0   +
         |
   67.5   +
         |
   65.0   +
         |
   62.5   +                                      N
         |
   60.0   +
         |
   57.5   +
         |
   55.0   +
         |                                        H
   52.5   +
         |       N
   50.0   +
         ---------+------------------------+------
               PLACEBO                  RITALIN

                        DRUG
```

The graph shows that normal children increase their activity when given ritalin while hyperactive children are calmed by ritalin. In the analysis of variance, the comparison of placebo to ritalin values is done by combining the data from normal and hyperactive children. Since these values tend to cancel each other, the average activity with placebo and ritalin is about the same. What we have found here is that it is **not** possible to understand the activity level of children by **just** knowing whether they had ritalin or not. One must also know whether they are hyperactive or not. This is why it is critical to understand the interaction before looking at main effects. If we really want to study the effect of ritalin, we should look **separately** at normal and hyperactive children. For each of these groups we have two levels of the DRUG. We can therefore do a t-test between placebo and ritalin within the normal and hyperactive groups. As we know from Chapter 4, this is accomplished by first sorting the data set by GROUP and then including a BY variable in the t-test request. We have

```
PROC SORT;
  BY GROUP;
PROC TTEST; BY GROUP;
  CLASS DRUG;
  VAR ACTIVITY;
```

The output from these statements is shown below:

```
GROUP=HYPER

TTEST PROCEDURE

VARIABLE: ACTIVITY

DRUG            N            MEAN            STD DEV          STD ERROR

PLACEBO         4        71.25000000      2.98607881        1.49303941
RITALIN         4        52.75000000      4.03112887        2.01556444

VARIANCES       T        DF   PROB > |T|

UNEQUAL       7.3755     5.5      0.0005
EQUAL         7.3755     6.0      0.0003

FOR HO: VARIANCES ARE EQUAL, F'=    1.82 WITH 3 AND 3 DF
        PROB > F'= 0.6343


GROUP=NORM -------------------------------------------------------------

VARIABLE: ACTIVITY

DRUG            N            MEAN            STD DEV          STD ERROR

PLACEBO         4        50.50000000      4.20317340        2.10158670
RITALIN         4        62.50000000      4.20317340        2.10158670

VARIANCES       T        DF   PROB > |T|

UNEQUAL      -4.0376     6.0      0.0068
EQUAL        -4.0376     6.0      0.0068

FOR HO: VARIANCES ARE EQUAL, F'=    1.00 WITH 3 AND 3 DF
        PROB > F'= 1.0000
```

Notice that in both groups, the two drug means are significantly different ($p < .05$). However, in the normal group, the ritalin mean is higher than the placebo mean while in the hyperactive group the reverse is true. So, **watch out for those interactions!**

An alternative to the t-tests above is to break down the two-way ANOVA into a one-way ANOVA by creating an independent (CLASS) variable that has a level for each **combination** of the original independent variables. In our case, we will create a variable (let's call it COND) that has a level for each combination of DRUG and GROUP. Thus, we will have NORMAL-PLACEBO, NORMAL-RITALIN, HYPERACTIVE-PLACEBO, and HYPERACTIVE-RITALIN as levels of our COND variable. A quick and easy way to create this variable is to **concatinate** (join) the two original independent variables (this will be performed directly for character variables. In the case of numeric variables, SAS software will convert them to character and perform the concatination). In the SAS system, concatination is accomplished with the concatination operator, "||".

To create the COND variable, we have

```
COND = GROUP || DRUG;
```

The SAS statements to produce the one-way ANOVA are

```
PROC ANOVA;
    TITLE 'ONE-WAY ANOVA RITALIN STUDY';
    CLASSES COND;
    MODEL ACTIVITY = COND;
    MEANS COND / DUNCAN;
```

The results of running this procedure are shown next:

```
Class Level Information

Class     Levels    Values

COND           4    HYPER  PLACEBO HYPER    RITALIN
                    NORM   PLACEBO NORM     RITALIN
```

```
Number of observations in data set = 16


Dependent Variable: ACTIVITY

Source                  DF  Sum of Squares     F Value      Pr > F

Model                    3  1093.50000000       24.10       0.0001

Error                   12   181.50000000

Corrected Total         15  1275.00000000

              R-Square              C.V.            ACTIVITY Mean

              0.857647          6.5638604            59.25000000


Source                  DF       Anova SS     F Value      Pr > F

COND                     3  1093.50000000       24.10       0.0001
```

Duncan's Multiple Range Test for variable: ACTIVITY

NOTE: This test controls the type I comparisonwise error rate,
      not the experimentwise error rate

Alpha= 0.05  df= 12  MSE= 15.125

```
Number of Means          2         3         4
Critical Range   5.9802927 6.2645621 6.4545514
```

Means with the same letter are not significantly different.

```
Duncan Grouping            Mean    N  COND

              A           71.250    4  HYPER   PLACEBO

              B           62.500    4  NORM    RITALIN

              C           52.750    4  HYPER   RITALIN
              C
              C           50.500    4  NORM    PLACEBO
```

Notice that this analysis tells us more than the two t-tests. Besides verifying that PLACEBO is different from RITALIN **within** each GROUP (NORMAL and HYPER), we can also see the other pairwise comparisons.

In the next chapter, dealing with repeated measures designs, we will find cases where it is the interaction term that is of primary interest.


## D. N-WAY FACTORIAL DESIGNS

The method we used to perform a two-way analysis of variance can be extended to cover any number of independent variables. An example with three independent variables (GROUP SEX DOSE) is shown below:

```
PROC ANOVA;
    TITLE '3-WAY ANALYSIS OF VARIANCE';
    CLASSES GROUP SEX DOSE;
    MEANS GROUP|SEX|DOSE;
    MODEL ACTIVITY = GROUP|SEX|DOSE;
```

With three independent variables, we have three main effects (GROUP SEX DOSE), three two-way interactions (GROUP*SEX GROUP*DOSE SEX*DOSE), and one three-way interaction (GROUP*SEX*DOSE). One usually hopes that the higher-order interactions are not significant since they complicate the interpretation of the main effects and the lower-order interactions. See Winer for a more complete discussion of this topic.

It clearly becomes difficult to perform factorial design experiments with a large number of independent variables without expert assistance. The number of subjects in the experiment will also need to be large so that there are a reasonable number of subjects per cell.


## E. UNBALANCED DESIGNS: PROC GLM

As we mentioned before, designs that have an unequal number of subjects per cell (called unbalanced designs) cannot be run using PROC ANOVA. PROC GLM (general linear model) is used instead. The only exception is for one-way unbalanced designs (designs with one CLASS variable and unequal numbers of subjects per group), which can be run with PROC ANOVA. CLASSES, MEANS, and MODEL statements for PROC GLM are identical to those used with PROC ANOVA. The only difference between the procedures is the mathematical methods used for each and some additional information that is computed when PROC GLM is used.

Here is an example of a two-way analysis of variance that is unbalanced:

A pudding company wanted to test-market a new product. Three levels of sweetness and two flavors were produced. Each subject was given a pudding to taste and was asked to rate the taste on a scale from 1 to 10. The following data were collected:

```
            SWEETNESS        LEVEL
             1        2        3
           -----------------------
             9        8        6
             7        7        5
VANILLA      8        8        7
             7
           -----------------------
             9        8        4
CHOCOLATE    9        7        5
             7        6        6
             7        8        4
             8                 4
```

The SAS INPUT statement was written:

| INPUT SWEET FLAVOR $ RATING;

Since the number of subjects in each cell is unequal, we will use PROC GLM.

```
PROC GLM;
    TITLE 'PUDDING TASTE EVALUATION';
    TITLE3 'TWO-WAY ANOVA - UNBALANCED DESIGN';
    TITLE5 '--------------------------------';
    CLASSES SWEET FLAVOR;
    MEANS SWEET|FLAVOR;
    MODEL RATING = SWEET|FLAVOR;
```

Portions of the output are shown below:

PUDDING TASTE EVALUATION

TWO-WAY ANOVA - UNBALANCED DESIGN

--------------------------------------

| SOURCE | DF | TYPE I SS | F VALUE | PR > F |
|---|---|---|---|---|
| SWEET | 2 | 35.85515873 | 21.00 | 0.0001 |
| FLAVOR | 1 | 1.33341530 | 1.56 | 0.2274 |
| SWEET*FLAVOR | 2 | 2.77809264 | 1.63 | 0.2241 |

| SOURCE | DF | TYPE III SS | F VALUE | PR > F |
|---|---|---|---|---|
| SWEET | 2 | 29.77706840 | 17.44 | 0.0001 |
| FLAVOR | 1 | 1.56666667 | 1.84 | 0.1923 |
| SWEET*FLAVOR | 2 | 2.77809264 | 1.63 | 0.2241 |

Notice that there are two sets of values for SUM OF SQUARES, F VALUES, and probabilities; one labeled TYPE I, the other labeled TYPE III. When designs do not have equal cell sizes, the TYPE I and TYPE III sums of squares may not be equal for all variables. The difference between TYPE I and TYPE III sum of squares is that TYPE I lists the sums of squares for each variable as if they were entered one at a time into the model in the order they are specified in the MODEL statement. Hence they can be thought of as incremental sums of squares. If there is any variance that is common to two or more variables, the variance will be attributed to the variable that is entered first. This may or may not be desirable. The TYPE III sum of squares gives the sum of squares that would be obtained for each variable if it were entered last into the model. That is, the effect of each variable is evaluated after all other factors have been accounted for. In any given situation, whether you want to look at TYPE I or TYPE III sum of squares will vary; however, for most analysis of variance applications, you will want to use TYPE III sum of squares.

Just to keep you on your toes, we have added a new form of the TITLE statement to the program. As you probably can guess, TITLE3 provides a third title line; TITLE5, a fifth. Since TITLE2 and TITLE4 are missing, lines 2 and 4 will be blank. In general, TITLEn will be the nth title line on the SAS output, where n is an integer. Note that TITLE is equivalent to TITLE1.

In our example, the sweetness factor was significant (p=.0001). The probabilities for FLAVOR and the interaction between FLAVOR and SWEETNESS were .1923 and 0.2241, respectively.

# Chapter 7    Repeated Measures Designs

**A BRIEF NOTE ABOUT REPEATED MEASURES**

Beginning with version 5 of SAS software, many of the repeated measures designs presented in this chapter can be programmed more quickly and easily (and run more cheaply) than with previous versions. The difference is the introduction of the **REPEATED** statement of ANOVA and GLM. However, some of the models presented here **cannot** take advantage of this new feature. Also, the older examples presented in this chapter that can be simplified are good exercises in teaching you about the SAS data step. Therefore, any examples that can take advantage of the **REPEATED** statement will do so but the original programming will be included as well.

**A. ONE-FACTOR EXPERIMENTS**

Consider the following experiment. We have four drugs (A,B,C,D) that relieve pain. Each subject is given each of the four drugs in random order. The subject's pain tolerance is then measured. Enough time is allowed to pass between successive drug administrations so that we can be sure there is no residual effect from the previous drug.

The null hypothesis is

MEAN(A)=MEAN(B)=MEAN(C)=MEAN(D)

If the analysis of variance is significant at p < .05 we will want to look at pairwise comparisons of the drugs using Duncan's multiple range test or other post hoc tests.

Notice how this experiment differs from a one-way analysis of variance without a repeated measure. With the designs we have discussed so far, we would have each subject receive only one of the four drugs. In this design, each subject is measured under each of the drug conditions. This has several important advantages.

First, each subject acts as his own control. That is, drug effects are calculated by recording deviations

between  each drug score and the average drug score  for
each  subject.  The normal subject-to-subject  variation
can thus be removed from the error sum of squares. Let's
look at a table of data from the pain experiment:

| SUBJECT | DRUG A | DRUG B | DRUG C | DRUG D |
|---------|--------|--------|--------|--------|
| 1 | 5 | 9 | 6 | 11 |
| 2 | 7 | 12 | 8 | 9 |
| 3 | 7 | 8 | 6 | 10 |
| 4 | 5 | 10 | 7 | 10 |

To  analyze this experiment,  we can  consider  the
subject  as an independent variable.  We therefore  have
SUBJECT and DRUG as independent variables.

One way of arranging our data and writing our INPUT
statement would be like this:

```
INPUT SUBJ DRUG PAIN;
CARDS;
1 1 5
1 2 9
1 3 6
1 4 11
2 1 7
 etc.
```

It  is  usually more convenient to arrange all  the
data for each subject on one line like this:

```
SUBJ  DRUG A  DRUG B  DRUG C  DRUG D
 1      5       9       6       11
 2      7       12      8       9
              etc.
```

We  can  read  the  data  arranged  like  this  but
restructure  it  to look as if we had read it  with  the
first program.

One way of writing the program is as follows:

```
DATA PAIN;
INPUT SUBJ @;
DO DRUG=1 TO 4;
   INPUT PAIN @;
   OUTPUT;
   END;
CARDS;
1 5 9 6 11
2 7 12 8 9
3 11 12 10 14
4 4 9 6 9
```

The first INPUT statement reads the subject number. The "@" sign following SUBJ is an instruction to keep reading from the same line of data. Normally, whenever a SAS program executes an INPUT statement, **it goes to a new line of data.** The next line, DO DRUG=1 to 4;, is an iterative loop. The value of DRUG is first set to 1. Next, the input statement, INPUT SUBJ @; is executed. Again, if the "@" were omitted, the program would go to the next data line to read a value (which we don't want). The OUTPUT statement causes an observation to be written to the internal SAS data set. Look at our first line of data. We would have as the first observation in the SAS data set SUBJ=1, DRUG=1, and PAIN=5. When the END statement is reached, the program flow returns to the DO statement where DRUG is set to 2. A new PAIN value is then read from the data (still from the first line because of the trailing @) and a new observation is added to the SAS data set. This continues until the value of DRUG=4. The general form of a DO statement is

```
DO variable = A TO B BY C;
   . . .

END;
```

Where A is the initial value for "variable," B is the ending value, and C is the increment. If C is omitted, the increment defaults to 1.

The initial value of "variable" will be set to A and incremented by C until the value of B is reached. Once the DO loop has been executed 4 times, we go on to the CARDS statement. This is normally the point at which observations are added to SAS data sets, but because we used an OUTPUT statement to write out our observation, no further action is taken. The program control returns to the statement INPUT PAIN @;. Since the CARDS statement was reached, the program will read the subject number from the next line of data. (Note: An input statement of the form **INPUT PAIN @@;** will allow the program to continue reading from the same line of data **even after the CARDS statement has been reached.**)

The first few observations in the SAS data set created from this program are listed below:

```
1 1 5
1 2 9
1 3 6
1 4 11
2 1 7
2 2 12
  etc.
```

We can make one small modification to the program and by so doing, avoid having to enter the subject numbers on each line of data. The new program looks as follows:

```
DATA PAIN;
SUBJ+1;
DO DRUG=1 TO 4;
   INPUT PAIN @;
   OUTPUT;
   END;
CARDS;
5 9 6 11
7 12 8 9
7 8 6 10
5 10 7 10
```

The statement SUBJ+1; creates a variable called SUBJ which starts at 1 and is incremented by 1 each time the statement is executed.

We are ready to write our PROC statements to analyze the data. With this design there are several ways to write the MODEL statement. One way is like this:

```
PROC ANOVA;
    CLASSES SUBJ DRUG;
    MEANS DRUG / DUNCAN;
    MODEL PAIN = SUBJ DRUG;
```

Notice that we are not writing SUBJ|DRUG. We are indicating that SUBJ and DRUG are each main effects and that there is no interaction term between them. Once we have accounted for variations from the grand mean due to subjects and drugs, the remaining deviations will be our source of error.

Below is a portion of the output from the one-way repeated measures experiment:

```
CLASS LEVEL INFORMATION

CLASS      LEVELS    VALUES

SUBJ          4      1 2 3 4

DRUG          4      1 2 3 4


NUMBER OF OBSERVATIONS IN DATA SET = 16
ANALYSIS OF VARIANCE PROCEDURE

DEPENDENT VARIABLE: PAIN
```

| SOURCE | DF | SUM OF SQUARES | MEAN SQUARE |
|---|---|---|---|
| MODEL | 6 | 132.50000000 | 22.08333333 |
| ERROR | 9 | 13.25000000 | 1.47222222 |

```
CORRECTED TOTAL              15          145.75000000

MODEL F =              15.00                              PR > F = 0.0003


R-SQUARE              C.V.              STD DEV              PAIN MEAN

0.909091              14.4878           1.21335165           8.37500000

SOURCE                       DF          ANOVA SS    F VALUE    PR > F

SUBJ                          3        82.25000000    18.62     0.0003
DRUG                          3        50.25000000    11.38     0.0020
```

DUNCAN'S MULTIPLE RANGE TEST FOR VARIABLE PAIN

MEANS WITH THE SAME LETTER ARE NOT SIGNIFICANTLY DIFFERENT.

ALPHA LEVEL=.05          DF=9          MS=1.47222

```
        GROUPING                  MEAN       N    DRUG

              A               10.250000      4     4
              A
              A               10.000000      4     2

              B                7.000000      4     3
              B
              B                6.250000      4     1
```

What conclusions can we draw from these results? Looking at the very bottom of the analysis of variance table, we find an F value of 11.38 with an associated probability of .0020. We can therefore reject the null hypothesis that the means are equal. Another way of saying this is that the four drugs are not equally effective in reducing pain. Notice that the SUBJ term in the analysis of variance table also has an F value and a probability associated with it. This merely tells us how much variability there was from subject to subject. It is not really interpretable in the same fashion as the drug factor. We include it as part of the model because we don't want the variability associated with it to go into the ERROR sum of squares.

Now that we know that the drugs are not equally effective, we can look at the results of the Duncan Multiple Range Test.  This shows two drug groupings. Assuming that a higher mean indicates higher pain, we can say that drugs 1 and 3 were more effective in reducing pain than drugs 2 and 4.  We cannot, at the .05 level, claim any differences between drugs 1 and 3 or between drugs 2 and 4.

Looking at the error SS and the SS due to subjects, we see that SUBJECT SS (82.25) is large compared to the ERROR SS (13.25).  Had this same set of data been the result of assigning 16 subjects to the 4 different drugs (instead of repeated measures), the error SS would have been 95.50 (13.25 + 82.25). The resulting F and p values for the DRUG effect would have been 2.10 and .1531, respectively.  (Note that the degrees of freedom for the error term would be 12 instead of 9.)

We see, therefore, that controlling for between-subject variability can greatly reduce the error term in our analysis of variance and allow us to identify small treatment differences with relatively few subjects.

## B. TWO-FACTOR EXPERIMENTS WITH A REPEATED MEASURE ON ONE FACTOR

One very popular form of a repeated measures design is the following:

```
                          PRE        POST
                  SUBJ
                   1
  CONTROL          2
                   3
-------------------------------------
                   4
  TREATMENT        5
                   6
```

Subjects are randomly assigned to a control or treatment group. Then, each subject is measured before and after treatment. The "treatment" for the control group can either be a placebo or no treatment at all. The goal of an experiment of this sort is to compare the pre/post changes of the control group to the pre/post changes of the treatment group. This design has a definite advantage over a simple pre/post design where one group of subjects is measured before and after a treatment (such as having only a treatment group in our design). Simple pre/post designs suffer from the problem that we cannot be sure if it is our treatment that causes a change (e.g., TIME may have an effect). By adding a pre/post control group, we can compare the pre/post control scores to the pre/post treatment scores and thereby control for any built in, systematic, pre/post changes.

A simple way to analyze our design is to compute a difference score (post minus pre) for each subject. We then have two groups of subjects with one score each (the difference score). Then we use a t-test to look for significant differences between the difference scores of the control and treatment groups.

Here are some sample data and a SAS program that calculates difference scores and computes a t-test:

|  | SUBJ | PRE | POST |
|---|---|---|---|
| CONTROL | 1 | 80 | 83 |
|  | 2 | 85 | 86 |
|  | 3 | 83 | 88 |
| TREATMENT | 4 | 82 | 94 |
|  | 5 | 87 | 93 |
|  | 6 | 84 | 98 |

```
DATA PREPOST;
INPUT SUBJ GROUP $ PRETEST POSTEST;
DIFF = POSTEST-PRETEST;
CARDS;
1 C 80 83
2 C 85 86
3 C 83 88
4 T 82 94
5 T 87 93
6 T 84 98
PROC TTEST;
    TITLE 'T-TEST ON DIFFERENCE SCORES';
    CLASS GROUP;
    VAR DIFF;
```

Results  of this analysis show the  treatment  mean difference to be  significantly  different from the control mean difference (p=.045). See below:

```
T-TEST ON DIFFERENCE SCORES

Ttest Procedure

Variable: DIFF

GROUP       N            Mean              Std Dev          Std Error
-------------------------------------------------------------------------
C           3         3.00000000         2.00000000        1.15470054
T           3        10.66666667         4.16333200        2.40370085

Variances        T        DF     Prob>|T|
---------------------------------------------
Unequal      -2.8750      2.9      0.0686
Equal        -2.8750      4.0      0.0452

For H0: Variances are equal, F'=    4.33 with 2 and 2 DF
        Prob > F'= 0.3750
```

We can alternatively treat this design as a two-way analysis  of  variance  (GROUP X TIME)  with TIME  as  a repeated  measure.  This  method has the  advantage  of analyzing  designs with more than two levels of  one  or both factors.

We will first write a program using the **REPEATED** statement of ANOVA. No changes in the data set are necessary. The ANOVA statements are

```
PROC ANOVA;
   CLASSES GROUP;
   TITLE 'TWO WAY ANOVA WITH A REPEATED MEASURE';
   TITLE2 'ON ONE FACTOR';
   MODEL PRETEST POSTEST = GROUP / NOUNI;
   REPEATED TIME   2 (0 1);
   MEANS GROUP;
```

The REPEATED statement indicates that we have a repeated measurement on a variable we have called TIME. The "2" following the variable name indicates that TIME has two levels. This is optional. Had it been omitted, the program would have assumed as many levels as there were dependent variables in the MODEL statement. The number of levels needs to be specified only when we have more than one repeated measure factor. Finally, "(0 1)" indicated the **labels** we want printed for each level of TIME. The labels also act as spacings when polynomial contrasts are requested. See the SAS/STAT manual for more details on this topic. The option "NOUNI" on the MODEL statement indicated that we do not want univariate statistics for each of the dependent variables on the MODEL statement.

Output from these procedure statements are shown below:

```
TWO-WAY ANOVA WITH A REPEATED MEASURE
ON ONE FACTOR

Analysis of Variance Procedure
Class Level Information

Class      Levels    Values

GROUP         2      C T
```

Number of observations in data set = 6


TWO-WAY ANOVA WITH A REPEATED MEASURE
ON ONE FACTOR

Analysis of Variance Procedure
Repeated Measures Analysis of Variance
Repeated Measures Level Information

Dependent Variable     PRETEST   POSTEST

     Level of TIME          0          1


Manova Test Criteria and Exact F Statistics for
the Hypothesis of no TIME Effect
H = Anova SS&CP Matrix for: TIME    E = Error SS&CP Matrix

$S=1$    $M=-0.5$    $N=1.5$

| Statistic | Value | F | Num DF | Den DF | Pr > F |
|---|---|---|---|---|---|
| Wilks' Lambda | .13216314 | 26.266 | 1 | 4 | 0.0069 |
| Pillai's Trace | .86783686 | 26.266 | 1 | 4 | 0.0069 |
| Hotelling-Lawley Trace | 6.5664063 | 26.266 | 1 | 4 | 0.0069 |
| Roy's Greatest Root | 6.5664063 | 26.266 | 1 | 4 | 0.0069 |


Manova Test Criteria and Exact F Statistics for
the Hypothesis of no TIME*GROUP Effect
H = Anova SS&CP Matrix for: TIME*GROUP    E = Error SS&CP Matrix

$S=1$    $M=-0.5$    $N=1.5$

| Statistic | Value | F | Num DF | Den DF | Pr > F |
|---|---|---|---|---|---|
| Wilks' Lambda | .32611465 | 8.2656 | 1 | 4 | 0.0452 |
| Pillai's Trace | .67388535 | 8.2656 | 1 | 4 | 0.0452 |
| Hotelling-Lawley Trace | 2.0664063 | 8.2656 | 1 | 4 | 0.0452 |
| Roy's Greatest Root | 2.0664063 | 8.2656 | 1 | 4 | 0.0452 |

```
TWO-WAY ANOVA WITH A REPEATED MEASURE
ON ONE FACTOR

Analysis of Variance Procedure
Repeated Measures Analysis of Variance
Tests of Hypotheses for Between Subjects Effects

Source                    DF        Anova SS      F Value     Pr > F

GROUP                     1       90.75000000      11.84      0.0263

Error                     4       30.66666667


TWO-WAY ANOVA WITH A REPEATED MEASURE                                  8
ON ONE FACTOR                        18:35 Tuesday, October 28, 1986

Analysis of Variance Procedure
Repeated Measures Analysis of Variance
Univariate Tests of Hypotheses for Within Subject Effects

Source: TIME
                                                       Adj  Pr > F
     DF    ANOVA SS Mean Square   F Value   Pr > F   G - G    H - F
      1  140.083333  140.083333    26.27    0.0069     .        .

Source: TIME*GROUP
                                                       Adj  Pr > F
     DF    ANOVA SS Mean Square   F Value   Pr > F   G - G    H - F
      1   44.083333   44.083333     8.27    0.0452     .        .

Source: ERROR(TIME)

     DF    ANOVA SS Mean Square
      4   21.333333    5.333333

Level of      ----------PRETEST----------  ----------POSTEST----------
GROUP    N     Mean           SD             Mean           SD

C        3   82.6666667    2.51661148     85.6666667    2.51661148
T        3   84.3333333    2.51661148     95.0000000    2.64575131
```

We will discuss the output from PROC ANOVA after we
show an alternative method of analyzing this experiment.
However, a portion of the output above is unique and
will be discussed here. Notice the rows labeled Wilks'
Lambda, Pillai's Trace, Hoteling-Lawley Trace, amd Roy's
Greatest Root. These are multivariate statistics which
are of special interest when more than one dependent
variable is indicated. Unlike univariate statistics,

when you use multivariate procedures, there is no single
test analogous to the F-test.   Instead,   there are about
half a dozen.   A question arises as to which one to use.
Multivariate statisticians spend many pleasant hours in-
vestigating this question. The answer to the question is
unambigiguous: it depends. Bock (1975--see Chapter 1 for
references)   presents a nice   discussion of the options.
Our   advice   is:   when there are only   very   small   differences   among the p-values,   it doesn't really   matter
which one you use. When there are differences   among the
p-values, find a consultant.

We will now analyze the same experiment as   a   two-
factor analysis of   variance **without** using the **REPEATED**
statement   of   PROC ANOVA.   To do this,   we   must   first
create   a   new variable--say TIME--which will   have   two
possible   values:   PRE or POST.   Each subject will   then
have two observations,   one with TIME = PRE and one with
TIME = POST.

As with our one-way,   repeated measures design, the
method of creating several observations from one is with
the OUTPUT statement.

We   can add the following SAS statements to the end
of the previous program:

```
1      DATA TWOWAY;
2      SET PREPOST;
3      TIME = 'PRE ';
4      SCORE = PRETEST;
5      OUTPUT;
6      TIME = 'POST';
7      SCORE = POSTEST;
8      OUTPUT;
9      DROP PRETEST POSTEST DIFF;
```

This section of the program creates a SAS data   set
called "TWOWAY"   which has variables SUBJ GROUP TIME and
SCORE. The first few observations in this data set are

```
SUBJ        GROUP       TIME        SCORE
 1            C          PRE          80
 1            C          POST         83
 2            C          PRE          85
 2            C          POST         86
```

Let's follow this portion of the SAS program step by step to see exactly how the new data set is created.

Line 1 prepares a data set called TWOWAY. Line 2 causes observations to be read from the original data set, PREPOST. The first observation is

SUBJ=1 GROUP=C PRETEST=80 POSTEST=83 DIFF=3.

Line 3 creates a new variable called TIME and sets the value of TIME to "PRE ." It should be noticed that there is a space after the 'E' in PRE. The reason is that the length of the variable TIME is defined by the first value that is assigned to it. Had we coded TIME = 'PRE', the length would be equal to three and the statement TIME = 'POST' would have assigned the value 'POS' to TIME instead of 'POST'.

Line 4 creates a new variable, SCORE, which is equal to the PRETEST value. When line 5 is executed, the first observation of the data set called TWOWAY becomes the following:

SUBJ=1 GROUP=C PRETEST=80 POSTEST=83 DIFF=3 TIME=PRE SCORE=80.

However, since we included a DROP statement in line 9, the first observation in data set TWOWAY is actually

SUBJ=1 GROUP=C TIME=PRE SCORE=80.

Next, line 6 sets TIME='POST' and line 7 sets the variable SCORE to the POSTEST value. A new observation is added to the data set TWOWAY in line 8. This second observation has

SUBJ=1 GROUP=C TIME=POST SCORE=83.

Finally,  as we mentioned, line 9 is an instruction to drop the variables PRETEST,  POSTEST,  and DIFF  from the  new data set.  Note:  The **DROP** statement can appear **anywhere** in the data  step  and  it controls which variables  get written to the SAS data set.  **Since the next line  of  the program is a SAS PROCEDURE,**  the  program logic returns to line 2 where a new observation is  read from the data set PREPOST.

We  are  now ready to write our  ANOVA  statements. Unlike  any of our previous examples,  we will  have  to specify all the terms,  including the sources of  error, in  the  MODEL statement.  This is necessary because our main effects and interaction terms are **not** tested by the same error term.  Therefore,  we need to specify each of these  terms in the MODEL statement so they can be  used later  in tests of our hypotheses.  In this design,  we have one group of  subjects that are  assigned to a control  group  and another group assigned to  a  treatment group.  Within each group,  each subject is measured at TIME=PRE and TIME=POST. In this design, the subjects are said to be **nested** within the GROUP. In SAS programs, the term subjects nested within group is written

SUBJ(GROUP)

Since  the model statement will define **ALL**  sources of  variation  about the grand mean,  the **ERROR  SUM  OF SQUARES** printed  in  the ANOVA table will be  **zero.**  To specify  which  error  term  to be  used  to  test  each hypothesis in  our design,  we will use **TEST**  statements following the MODEL specification. A TEST statement consists  of  a hypothesis to be tested  (H=)  and the appropriate  error term (E=).  The entire ANOVA  procedure looks as follows:

```
PROC ANOVA;
   TITLE '2 WAY ANOVA WITH TIME AS A REPEATED MEASURE';
   CLASSES SUBJ GROUP TIME;
   MEANS GROUP|TIME;
   MODEL SCORE = GROUP SUBJ(GROUP) TIME
                 GROUP*TIME TIME*SUBJ(GROUP);
   TEST H=GROUP             E=SUBJ(GROUP);
   TEST H=TIME GROUP*TIME   E=TIME*SUBJ(GROUP);
```

Notice that the error term for GROUP is SUBJ(GROUP) (subject nested within group) and the error term for TIME and the GROUP*TIME interaction is TIME*SUBJ(GROUP).

Below are portions of the PROC ANOVA output:

```
CLASS LEVEL INFORMATION

CLASS      LEVELS    VALUES

SUBJ          6      1 2 3 4 5 6

GROUP         2      C T

TIME          2      POST PRE

DEPENDENT VARIABLE: SCORE

SOURCE              DF      SUM OF SQUARES        MEAN SQUARE

MODEL               11       326.91666667        29.71969697

ERROR                0         0.00000000         0.00000000

CORRECTED TOTAL     11       326.91666667


MODEL F =      99999.99                     PR > F = 0.0000


SOURCE              DF           ANOVA SS    F VALUE     PR > F

GROUP                1         90.75000000      .          .
SUBJ(GROUP)          4         30.66666667      .          .
TIME                 1        140.08333333      .          .
GROUP*TIME           1         44.08333333      .          .
SUBJ*TIME(GROUP)     4         21.33333333      .          .
```

```
TESTS OF HYPOTHESES USING THE ANOVA MS FOR SUBJ(GROUP)
AS AN ERROR TERM

SOURCE                    DF           ANOVA SS    F VALUE    PR > F

GROUP                      1          90.75000000    11.84    0.0263


TESTS OF HYPOTHESES USING THE ANOVA MS FOR SUBJ*TIME(GROUP)
AS AN ERROR TERM

SOURCE                    DF           ANOVA SS    F VALUE    PR > F

TIME                       1         140.08333333    26.27    0.0069
GROUP*TIME                 1          44.08333333     8.27    0.0452
```

Since all sources of variation were included in the MODEL statement, the error sum of squares is zero and the F value is undefined (it prints as 99999.99). The requested tests are shown at the bottom of the table. Group differences had an $F=11.84$ and $p=.0263$. TIME and GROUP*TIME had F values of 26.27 and 8.27 and probabilities of .0069 and .0452, respectively.

In this experimental design, it is the interaction of GROUP and TIME that is of primary importance. This interaction term tells us **if the pre/post changes were the same for control and treatment subjects.** An interaction graph will make this clear. The output from the MEANS request is shown below:

```
MEANS

GROUP         N         SCORE

C             6       84.1666667
T             6       89.6666667


TIME          N         SCORE

POST          6       90.3333333
PRE           6       83.5000000
```

| GROUP | TIME | N | SCORE |
|-------|------|---|-------|
| C | POST | 3 | 85.6666667 |
| C | PRE | 3 | 82.6666667 |
| T | POST | 3 | 95.0000000 |
| T | PRE | 3 | 84.3333333 |

We can use the last set of means (interaction of GROUP and TIME) to plot the interaction graph. We pick one of the independent variables (we will use TIME) to go on the x-axis and then plot means for each of the levels of the other independent variable (GROUP). The resulting graph is shown below:

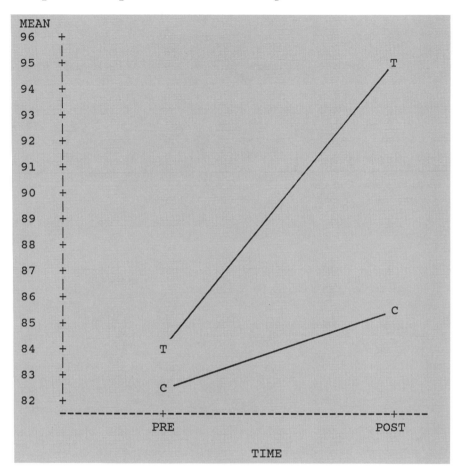

```
MEAN
96    +
      |
95    +                                                    T
      |
94    +
      |
93    +
      |
92    +
      |
91    +
      |
90    +
      |
89    +
      |
88    +
      |
87    +
      |
86    +
      |                                                    C
85    +
      |
84    +        T
      |
83    +
      |        C
82    +
      -------------+---------------------------------+----
                  PRE                              POST

                          TIME
```

A significant interaction term shows us that the
two pre/post lines are not parallel. This tells us that
the change from pre to post was different depending on
which GROUP a subject was in. This is precisely what we
wanted to know. The treatment group and control group
were quite similar in terms of pain relief **before** the
drug was administered (mean=84.33 and 82.67). After the
drug was given (the POST measure), the treatment group
showed dramatic gains and the control group only modest
gains. The F statistic for GROUP X TIME (8.27) and its
p-value (.045) tell us that this difference in improvement is greater than could be expected by chance alone.

The F statistic for GROUP ($F=11.84$, $p=.0263$) tells us that if we summed over the pre and post tests, the groups were different. This isn't of use to us since it combines the pre measure (where we anticipated them being the same) with the post measure (where we anticipated a difference). The same logic is true for TIME. Here we are summing over the control and treatment groups. Finally, note that the p-value for GROUP X TIME is **the same** as for the t-test of the difference scores. This is because we are essentially making the same test in both analyses. Next, we will move to a somewhat more complex setting.

## C. TWO-FACTOR EXPERIMENTS WITH REPEATED MEASURES ON BOTH FACTORS

This design is similar to the previous design except that each subject is measured under all levels of both factors. Note that since **both** factors are repeated (i.e., there are no crossed factors), we **cannot** use the REPEATED statement of PROC ANOVA or GLM. (There would be no dependent variable to put on the CLASSES statement or the MODEL statement.) An example follows.

A group of subjects is tested in the morning and afternoon of two different days. On one of the days, the subjects receive a strong sleeping aid the night before the experiment is to be conducted, on the other, a placebo. The subjects' reaction time to a stimulus is measured. A diagram of the experiment is shown below:

```
                            TREAT

              CONTROL                      DRUG
   TIME    ---------------------------------------------
          | SUBJ   REACT            SUBJ   REACT|
          |   1      65                1      70  |
   A.M.   |   2      72                2      78  |
          |   3      90                3      97  |
   --------------------------------------------------|
          |   1      55                1      60  |
   P.M.   |   2      64                2      68  |
          |   3      80                3      85  |
           ---------------------------------------------
```

We would like to see if the drug had any effect on the reaction time and if the effect was the same for the whole day. We can use the AM/PM measurements on the control day as a comparison for the AM/PM changes on the drug day.

Since each subject is measured under all levels of treatment (PLACEBO or DRUG) and TIME (AM/PM), we can treat this experiment as a SUBJ by TREATMENT by TIME factorial design. However, we must specify the error terms to test our hypotheses.

To create our SAS data set, we could use the following statements:

```
DATA SLEEP;
INPUT SUBJ TREAT $ TIME $ REACT;
CARDS;
1 CONT AM 65
1 CONT PM 55
1 DRUG AM 70
1 DRUG PM 60
2 CONT AM 72
   etc.
```

The ANOVA statements can be written:

```
PROC ANOVA;
   CLASSES SUBJ TREAT TIME;
   MODEL REACT = SUBJ|TREAT|TIME;
   MEANS TREAT|TIME;
   TEST H=TREAT        E=SUBJ*TREAT;
   TEST H=TIME         E=SUBJ*TIME;
   TEST H=TREAT*TIME   E=SUBJ*TREAT*TIME;
```

Before we investigate the output from the above program, we would like to show an alternate way of programming this problem. This method differs from the one above only in the way that data are read; the PROC statements are exactly the same. The purpose of the alternate programming method is to simplify data entry. Please feel free to skip (to below the solid line) this discussion if you wish; it is useful only if you will be using SAS software frequently with moderate to large amounts of data. In that case you will save considerable time. Here is the program:

ALTERNATIVE PROGRAM FOR SLEEP STUDY

```
1       DATA;
2       SUBJ+1;
3       DO TIME=1 TO 2;
4           DO TREAT=1 TO 2;
5           INPUT REACT @;
6           OUTPUT;
7           END;
8       END;
9       CARDS;
10      65 70 55 60
11      72 78 64 68
12      90 97 80 85

13      PROC ANOVA;
            etc.
```

This program allows us to place all the data for one subject on a single line. We begin creating our data set with the DATA statement on line 1. Since we are not explicitly entering a subject number, line 2 will provide us with a SUBJ variable that starts with 1 and is incremented by 1 for each new subject.

The reaction times for each subject are arranged as follows:

CONTROL AM  -  DRUG AM  -  CONTROL PM  -  DRUG PM

We want to create 4 observations for each subject (one for each combination of treatment and time). Line 3 first sets the variable TIME equal to 1. Then the inner "DO" loop (lines 4 through 7) first sets TREAT equal to 1. The first response time (65) is read and an observation is placed into the data set with the OUTPUT statement (line 6). The first observation is therefore

SUBJ=1 TIME=1 TREAT=1 REACT=65

The inner loop now sets TREAT=2, reads the next reaction time (on the same line because of the trailing

'@' sign in the INPUT statement), and outputs the next observation:

SUBJ=1 TIME=1 TREAT=2 REACT=70

Since the inner loop has reached its limit (2), the program logic returns to the outer loop starting at line 3, where the value of TIME is set to 2. The inner loop then outputs two more observations (TREAT=1 and 2). At this point, we return to line 2 where SUBJ is incremented by 1 and four more observations will be added to the data set.

A **FORMAT** statement to assign formats to the variables TREAT and TIME would make output from the statistical procedures easier to read. The complete program, modified to include formats, is shown next:

```
PROC FORMAT;
    VALUE FTREAT 1=CONTROL 2=DRUG;
    VALUE FTIME 1=AM 2=PM;
DATA;
SUBJ+1;
DO TIME=1 TO 2;
    DO TREAT=1 TO 2;
    INPUT REACT @;
    OUTPUT;
    END;
END;
FORMAT TREAT FTREAT. TIME FTIME.;
CARDS;
65 70 55 60
72 78 64 68
90 97 80 85

PROC ANOVA;
    etc.
```

**END OF ALTERNATIVE PROGRAM EXPLANATION**

Which method you choose to create the SAS data set will not affect the PROC ANOVA statements. In any design where **ALL** factors are repeated, such as this one, we can treat the SUBJ variable as being crossed by all other factors (as opposed to nested). The MODEL statement is therefore the same as our factorial design. However, by including the SUBJ term in our model, the error term will be zero (as in our previous example). Thus, our ANOVA table will **not** show F values or probabilities. These will be obtained by specifying **TEST** statements following the MODEL statement as previously described.

The error terms to test each hypothesis are simple to remember: For factor X, the error term will be SUBJ*X. For example, the error term to test TREAT is SUBJ*TREAT; the error term to test the interaction TREAT*TIME is SUBJ*TREAT*TIME. To specify the correct error term for each main effect and interaction, the three TEST statements following the MODEL statement were added, each specifying a hypothesis to be tested and the error term to be used in calculating the F ratio.

A portion of the output from PROC ANOVA is shown below:

ANALYSIS OF VARIANCE PROCEDURE

DEPENDENT VARIABLE: REACT

| SOURCE | DF | SUM OF SQUARES | MEAN SQUARE |
|---|---|---|---|
| MODEL | 11 | 1750.66666667 | 159.15151515 |
| ERROR | 0 | 0.00000000 | 0.00000000 |
| CORRECTED TOTAL | 11 | 1750.66666667 | |

| SOURCE | DF | ANOVA SS | F VALUE | PR > F |
|---|---|---|---|---|
| SUBJ | 2 | 1360.66666667 | . | . |
| TREAT | 1 | 85.33333333 | . | . |
| SUBJ*TREAT | 2 | 0.66666667 | . | . |
| TIME | 1 | 300.00000000 | . | . |
| SUBJ*TIME | 2 | 2.00000000 | . | . |
| TREAT*TIME | 1 | 1.33333333 | . | . |
| SUBJ*TREAT*TIME | 2 | 0.66666667 | . | . |

```
TESTS OF HYPOTHESES USING THE ANOVA MS FOR SUBJ*TREAT
AS AN ERROR TERM

SOURCE                  DF              ANOVA SS      F VALUE     PR > F

TREAT                   1           85.33333333      256.00     0.0039

TESTS OF HYPOTHESES USING THE ANOVA MS FOR SUBJ*TIME
AS AN ERROR TERM

SOURCE                  DF              ANOVA SS      F VALUE     PR > F

TIME                    1          300.00000000      300.00     0.0033

TESTS OF HYPOTHESES USING THE ANOVA MS FOR SUBJ*TREAT*TIME
AS AN ERROR TERM

SOURCE                  DF              ANOVA SS      F VALUE     PR > F

TREAT*TIME              1            1.33333333        4.00     0.1835
```

What conclusions can we draw? (1) The drug increases reaction time (F=256.00, p=.0039), (2) reaction time is longer in the morning compared to the afternoon (F=300.00, p=.0033), and (3) we cannot conclude that the effect of the drug on reaction time is related to the time of day (the interaction of TREAT and TIME is not significant F=4.00, p=0.1835). Note that this study was **not** a pre/post study as in the previous example. Even so, had the TREAT X TIME interaction been significant, we would have been more cautious in looking at the TREAT and TIME effects.

If this were a real experiment, we would have to control for the learning effect that might take place. For example, if we measure each subject in the same order, we might find a decrease in reaction time from CONTROL AM to DRUG PM because the subject became more familiar with the apparatus. To avoid this, we would either have to acquaint the subject with the equipment before the experiment begins to assure ourselves that the learning has stabilized, or to measure each subject using TREATMENT and TIME in random order.

## D. THREE-FACTOR EXPERIMENTS WITH A REPEATED MEASURE ON THE LAST FACTOR

For this example, we will consider a marketing experiment. Male and female subjects are offered one of three different brands of coffee. Each brand is tasted twice; once immediately after breakfast, the other time after dinner (the order of presentation is randomized for each subject). The preference of each brand is measured on a scale from 1 to 10 (1=lowest, 10=highest). The experimental design is shown below:

```
                          BRAND (of coffee)
                  A                   B                    C
            BRKFST DINNER       BRKFST DINNER       BRKFST DINNER
    S
    E       -------------------------------------------------------
    X    subj              |subj               |subj
            1     7     8   |  7     4     6    |  13    8     9
   MALE  2     6     7   |  8     3     5    |  14    6     9
            3     6     8   |  9     3     5    |  15    5     8
         -------------------------------------------------------
            4     5     7   | 10     3     4    |  16    6     9
   FEM   5     4     7   | 11     4     4    |  17    6     9
            6     4     6   | 12     2     3    |  18    7     8
```

In this experiment, the factors BRAND and SEX are crossed factors while MEAL is a repeated measure factor (each subject tastes coffee after breakfast and dinner). Since a single subject tastes only **one** brand of coffee and is clearly only **one** sex, the subject term is said to be nested within BRAND and SEX (written SUBJ(BRAND SEX)). We could arrange our data several ways. First, we will arrange data so that we can take advantage of the **REPEATED** statement of ANOVA. To do this, we will place all data for each subject on one line. Thus, our program and data will look as follows:

```
DATA;
INPUT SUBJ BRAND $ SEX $ SCORE_B SCORE_D;
CARDS;
1 A M 7 8
2 A M 6 7
3 A M 6 8
4 A F 5 7
5 A F 4 7
6 A F 4 6
7 B M 4 6
   etc.
PROC ANOVA;
   TITLE 'COFFEE STUDY';
   CLASSES BRAND SEX;
   MODEL SCORE_B SCORE_D = BRAND|SEX / NOUNI;
   REPEATED MEAL;
   MEANS BRAND|SEX;
```

Notice that BRAND and SEX are crossed while MEAL is the repeated measures factor. As before, the option NOUNI on the MODEL statement indicates that we do **not** want **UNI**variate statistics for SCORE_B and SCORE_D.

Selected portions of the output from the above program are shown below:

```
COFFEE STUDY

Analysis of Variance Procedure
Class Level Information

Class     Levels     Values

BRAND          3     A B C

SEX            2     F M


Number of observations in data set = 18

Repeated Measures Analysis of Variance
Repeated Measures Level Information

Dependent Variable     SCORE_B   SCORE_D

   Level of MEAL           1         2
```

Tests of Hypotheses for Between Subjects Effects

| Source | DF | Anova SS | Mean Square | F Value | Pr > F |
|---|---|---|---|---|---|
| BRAND | 2 | 83.388889 | 41.694444 | 51.76 | 0.0001 |
| SEX | 1 | 6.250000 | 6.250000 | 7.76 | 0.0165 |
| BRAND*SEX | 2 | 3.500000 | 1.750000 | 2.17 | 0.1566 |
| Error | 12 | 9.666667 | 0.805556 | | |

Univariate Tests of Hypotheses for Within Subject Effects

Source: MEAL

| DF | ANOVA SS | Mean Square | F Value | Pr > F |
|---|---|---|---|---|
| 1 | 30.25000000 | 30.25000000 | 99.00 | 0.0001 |

Source: MEAL*BRAND

| DF | ANOVA SS | Mean Square | F Value | Pr > F |
|---|---|---|---|---|
| 2 | 1.50000000 | 0.75000000 | 2.45 | 0.1278 |

Source: MEAL*SEX

| DF | ANOVA SS | Mean Square | F Value | Pr > F |
|---|---|---|---|---|
| 1 | 0.02777778 | 0.02777778 | 0.09 | 0.7682 |

Source: MEAL*BRAND*SEX

| DF | ANOVA SS | Mean Square | F Value | Pr > F |
|---|---|---|---|---|
| 2 | 2.05555556 | 1.02777778 | 3.36 | 0.0692 |

Source: ERROR(MEAL)

| DF | ANOVA SS | Mean Square |
|---|---|---|
| 12 | 3.66666667 | 0.30555556 |

| Level of BRAND | N | --------SCORE_B--------- Mean | SD | --------SCORE_D-------- Mean | SD |
|---|---|---|---|---|---|
| A | 6 | 5.33333333 | 1.21106014 | 7.16666667 | 0.75277265 |
| B | 6 | 3.16666667 | 0.75277265 | 4.50000000 | 1.04880885 |
| C | 6 | 6.33333333 | 1.03279556 | 8.66666667 | 0.51639778 |

| Level of SEX | N | -------SCORE_B--------- Mean | SD | --------SCORE_D-------- Mean | SD |
|---|---|---|---|---|---|
| F | 9 | 4.55555556 | 1.58989867 | 6.33333333 | 2.23606798 |
| M | 9 | 5.33333333 | 1.73205081 | 7.22222222 | 1.56347192 |

| Level of | Level of | | ------SCORE_B------ | | ------SCORE_D------ | |
|---|---|---|---|---|---|---|
| BRAND | SEX | N | Mean | SD | Mean | SD |
| A | F | 3 | 4.333333 | 0.577350 | 6.666666 | 0.577350 |
| A | M | 3 | 6.333333 | 0.577350 | 7.666666 | 0.577350 |
| B | F | 3 | 3.000000 | 1.000000 | 3.666666 | 0.577350 |
| B | M | 3 | 3.333333 | 0.577350 | 5.333333 | 0.577350 |
| C | F | 3 | 6.333333 | 0.577350 | 8.666666 | 0.577350 |
| C | M | 3 | 6.333333 | 1.527525 | 8.666666 | 0.577350 |

We will explain the results after the alternative program below:

An alternative program can be written that does not use the REPEATED statement of PROC ANOVA. You might find this useful if the data were arranged with two observations per subject and a MEAL variable already in the data set. So, if the data were arranged like this:

```
SUBJ BRAND SEX   MEAL   SCORE
-----------------------------
  1    A    M   BRKFST    7
  1    A    M   DINNER    8
  2    A    M   BRKFST    6
          etc.
```

your INPUT statement would look like this:

```
INPUT  SUBJ BRAND $ SEX $ MEAL $ SCORE;
```

The ANOVA statements are written

```
PROC ANOVA;
   CLASSES SUBJ BRAND SEX MEAL;
   MODEL SCORE = BRAND SEX BRAND*SEX SUBJ(BRAND SEX)
                 MEAL BRAND*MEAL SEX*MEAL BRAND*SEX*MEAL
                 MEAL*SUBJ(BRAND SEX);
   MEANS BRAND|SEX / DUNCAN E=SUBJ(BRAND SEX);
   MEANS MEAL BRAND*MEAL SEX*MEAL BRAND*SEX*MEAL;
```

This is followed by TEST statements:

```
TEST H=BRAND SEX BRAND*SEX
     E=SUBJ(BRAND SEX);
TEST H=MEAL BRAND*MEAL SEX*MEAL BRAND*SEX*MEAL
     E=MEAL*SUBJ(BRAND SEX);
```

The first test statement will test **each** of the terms (BRAND SEX and BRAND*SEX) with the error term SUBJ(BRAND SEX). The effects MEAL, BRAND*MEAL, SEX*MEAL, and BRAND*SEX*MEAL will all be tested with the error term MEAL*SUBJ(BRAND SEX). We have also made a change in the way the MEANS statements were written. Included after the DUNCAN option is an "E=" specification. This is done because the DUNCAN procedure will use the residual mean square as the error term unless otherwise instructed. Since we have completely defined every source of variation in our model, the residual mean square is zero. The "E=error term" option uses the same error term as the "H=" option of the corresponding TEST statement. Also, since different error terms are used to test different hypotheses, a separate MEANS statement is required each time a different error term is used. Note then we did not need to perform a DUNCAN test for MEAL since this variable has only two levels.

A portion of the results of running this alternative program are shown below:

```
Analysis of Variance Procedure Class Level Information

Class     Levels    Values

SUBJ         18      1 2 3 4 5 6 7 8 9 10 11 12 13 14 15 16 17 18

BRAND         3      A B C

SEX           2      F M

MEAL          2      BRKFST DINNER


Number of observations in data set = 36

Source                DF    Anova SS   Mean Square   F Value   Pr > F

BRAND                  2   83.388889   41.694444        .         .
SEX                    1    6.250000    6.250000        .         .
BRAND*SEX              2    3.500000    1.750000        .         .
SUBJ(BRAND*SEX)       12    9.666667    0.805556        .         .
MEAL                   1   30.250000   30.250000        .         .
BRAND*MEAL             2    1.500000    0.750000        .         .
SEX*MEAL               1    0.027778    0.027778        .         .
BRAND*SEX*MEAL         2    2.055556    1.027778        .         .
SUBJ*MEAL(BRAND*SEX)  12    3.666667    0.305556        .         .
```

```
Duncan's Multiple Range Test for variable: SCORE

NOTE: This test controls the type I comparisonwise error rate,
      not the experimentwise error rate

Alpha= 0.05  df= 12  MSE= .8055556

Number of Means           2          3
Critical Range   .79682302 .83469949

Means with the same letter are not significantly different.

Duncan Grouping              Mean      N  BRAND

                 A           7.500     12  C

                 B           6.250     12  A

                 C           3.833     12  B


Level of  Level of        --------------SCORE-------------
BRAND     SEX       N      Mean                   SD

A         F         6      5.50000000             1.37840488
A         M         6      7.00000000             0.89442719
B         F         6      3.33333333             0.81649658
B         M         6      4.33333333             1.21106014
C         F         6      7.50000000             1.37840488
C         M         6      7.50000000             1.64316767

Level of               --------------SCORE-------------
MEAL          N        Mean                   SD

BRKFST        18       4.94444444             1.66175748
DINNER        18       6.77777778             1.92676369

Level of            --------------SCORE-------------
BRAND     N         Mean                   SD

A         12        6.25000000             1.35680105
B         12        3.83333333             1.11464086
C         12        7.50000000             1.44599761
```

| Level of BRAND | Level of MEAL | N | --------------SCORE-------------- Mean | SD |
|---|---|---|---|---|
| A | BRKFST | 6 | 5.33333333 | 1.21106014 |
| A | DINNER | 6 | 7.16666667 | 0.75277265 |
| B | BRKFST | 6 | 3.16666667 | 0.75277265 |
| B | DINNER | 6 | 4.50000000 | 1.04880885 |
| C | BRKFST | 6 | 6.33333333 | 1.03279556 |
| C | DINNER | 6 | 8.66666667 | 0.51639778 |

Tests of Hypotheses using the Anova MS for SUBJ(BRAND*SEX) as an error term

| Source | DF | Anova SS | Mean Square | F Value | Pr > F |
|---|---|---|---|---|---|
| BRAND | 2 | 83.38888889 | 41.69444444 | 51.76 | 0.0001 |
| SEX | 1 | 6.25000000 | 6.25000000 | 7.76 | 0.0165 |
| BRAND*SEX | 2 | 3.50000000 | 1.75000000 | 2.17 | 0.1566 |

Tests of Hypotheses using the Anova MS for SUBJ*MEAL(BRAND*SEX) as an error term

| Source | DF | Anova SS | Mean Square | F Value | Pr > F |
|---|---|---|---|---|---|
| MEAL | 1 | 30.25000000 | 30.25000000 | 99.00 | 0.0001 |
| BRAND*MEAL | 2 | 1.50000000 | 0.75000000 | 2.45 | 0.1278 |
| SEX*MEAL | 1 | 0.02777778 | 0.02777778 | 0.09 | 0.7682 |
| BRAND*SEX*MEAL | 2 | 2.05555556 | 1.02777778 | 3.36 | 0.0692 |

What conclusions can we draw from these results? First, we notice that the variables BRAND, MEAL, and SEX are all significant effects (BRAND and MEAL at p=.0001, SEX at p=.016). We see, from the Duncan test, that brand C is the preferred brand, followed by A and B. The fact that MEAL (breakfast or dinner) is significant and that BRAND*MEAL is not, tells us that all three brands of coffee are preferred after dinner.

## E. THREE-FACTOR EXPERIMENTS WITH REPEATED MEASURES ON TWO FACTORS

As an example of a three-factor experiment with two repeated measures factors, we have designed a hypothetical study involving reading comprehension and a concept called "slippage." It is well known that many students will do less well on a reading comprehension test in the early fall compared to the previous spring because of "slippage" during the summer vacation. As children grow older, the slippage should decrease. Also, slippage tends to be smaller with high-SES (socio-economic status--roughly speaking, "wealthier") children compared to low-SES children, since high-SES children typically do more reading over the summer.

To test these ideas, the following experiment was devised:

A group of high- and low-SES children was selected for the experiment. Their reading comprehension was tested each spring and fall for three consecutive years. A diagram of the design is shown below:

|  |  | YEARS 1 | | 2 | | 3 | |
|---|---|---|---|---|---|---|---|
|  |  | SPRING | FALL | SPRING | FALL | SPRING | FALL |
| HIGH | subj |  |  |  |  |  |  |
| SES | 1 | 61 | 50 | 60 | 55 | 59 | 62 |
|  | 2 | 64 | 55 | 62 | 57 | 63 | 63 |
|  | 3 | 59 | 49 | 58 | 52 | 60 | 58 |
|  | 4 | 63 | 59 | 65 | 64 | 67 | 70 |
|  | 5 | 62 | 51 | 61 | 56 | 60 | 63 |
| LOW | 6 | 57 | 42 | 56 | 46 | 54 | 50 |
| SES | 7 | 61 | 47 | 58 | 48 | 59 | 55 |
|  | 8 | 55 | 40 | 55 | 46 | 57 | 52 |
|  | 9 | 59 | 44 | 61 | 50 | 63 | 60 |
|  | 10 | 58 | 44 | 56 | 49 | 55 | 49 |

Notice that each subject is measured each spring and fall and each year so that the variables SEASON and YEAR are both repeated measures factors. In this design each subject belongs to either the high-SES or the low-SES group. Therefore, subjects are **nested** within SES.

We will show three ways of writing a SAS program to analyze this experiment. First, using the REPEATED statement of PROC ANOVA:

```
DATA;
INPUT SUBJ SES $ READ1-READ6;
LABEL READ1 = 'SPRING YR 1'
      READ2 = 'FALL YR 1'
      READ3 = 'SPRING YR 2'
      READ4 = 'FALL YR 2'
      READ5 = 'SPRING YR 3'
      READ6 = 'FALL YR 3';
CARDS;
1 HIGH 61 50 60 55 59 62
2 HIGH 64 55 62 57 63 63
3 HIGH 59 49 58 52 60 58
4 HIGH 63 59 65 64 67 70
5 HIGH 62 51 61 56 60 63
6 LOW  57 42 56 46 54 50
7 LOW  61 47 58 48 59 55
8 LOW  55 40 55 46 57 52
9 LOW  59 44 61 50 63 60
10 LOW 58 44 56 49 55 49
PROC ANOVA;
   CLASSES SES;
   MODEL READ1-READ6 = SES / NOUNI;
   REPEATED YEAR 3, SEASON 2;
   MEANS SES;
```

Since this is the first time we have had two repeated measures factors, a word of explanation is in order. The data are arranged in the order

```
       YEAR 1           YEAR 2           YEAR 3
   SPRING  FALL     SPRING  FALL     SPRING  FALL
      1      2         3      4         5      6
```

There are **three** levels of YEAR and **two** levels of SEASON. The **factors** following the keyword REPEATED are placed in order from the one that varies the **slowest** to the one that varies the **fastest**. For example, the first number (READ1) is from YEAR 1 in the SPRING. The next number (READ2) is still YEAR 1 but in the FALL. Thus, we say that SEASON is varying faster than YEAR. We must also be sure to indicate the number of levels of each

factor following the factor  name on the REPEATED statement.

"REPEATED YEAR 3, SEASON 2;" tells the SAS software
to choose the first level of YEAR (1), then loop through
two levels of SEASON (SPRING FALL),  then return to  the
next  level  of  YEAR (2),  followed by  two  levels  of
SEASON,   etc.  The product of the two levels must equal
the  number of variables in the dependent variable  list
of  the MODEL statement.  To check,  3*2=6  and we  have
READ1 to READ6 on the MODEL statement.

Results of running this program are shown next:

```
Analysis of Variance Procedure
Class Level Information

Class     Levels    Values

SES         2       HIGH LOW


Number of observations in data set = 10


Repeated Measures Analysis of Variance
Repeated Measures Level Information

Dependent Variable     READ1   READ2   READ3   READ4   READ5  READ6

    Level of YEAR        1       1       2       2       3      3
    Level of SEASON      1       2       1       2       1      2

Tests of Hypotheses for Between Subjects Effects

Source          DF     Anova SS  Mean Square    F Value    Pr > F

SES              1     680.0667    680.0667      13.54     0.0062

Error            8     401.6667     50.2083
```

Univariate Tests of Hypotheses for Within Subject Effects

Source: YEAR

| DF | ANOVA SS | Mean Square | F Value | Pr > F | Adj Pr > F G - G | H - F |
|---|---|---|---|---|---|---|
| 2 | 252.03333333 | 126.01666667 | 26.91 | 0.0001 | 0.0002 | 0.0001 |

Source: YEAR*SES

| DF | ANOVA SS | Mean Square | F Value | Pr > F | Adj Pr > F G - G | H - F |
|---|---|---|---|---|---|---|
| 2 | 1.03333333 | 0.51666667 | 0.11 | 0.8962 | 0.8186 | 0.8700 |

Source: ERROR(YEAR)

| DF | ANOVA SS | Mean Square |
|---|---|---|
| 16 | 74.93333333 | 4.68333333 |

Greenhouse-Geisser Epsilon = 0.6757
Huynh-Feldt Epsilon = 0.8658


Source: SEASON

| DF | ANOVA SS | Mean Square | F Value | Pr > F |
|---|---|---|---|---|
| 1 | 680.06666667 | 680.06666667 | 224.82 | 0.0001 |

Source: SEASON*SES

| DF | ANOVA SS | Mean Square | F Value | Pr > F |
|---|---|---|---|---|
| 1 | 112.06666667 | 112.06666667 | 37.05 | 0.0003 |

Source: ERROR(SEASON)

| DF | ANOVA SS | Mean Square |
|---|---|---|
| 8 | 24.20000000 | 3.02500000 |

Source: YEAR*SEASON

| DF | ANOVA SS | Mean Square | F Value | Pr > F |
|---|---|---|---|---|
| 2 | 265.43333333 | 132.71666667 | 112.95 | 0.0001 |

Source: YEAR*SEASON*SES

| DF | ANOVA SS | Mean Square | F Value | Pr > F |
|---|---|---|---|---|
| 2 | 0.43333333 | 0.21666667 | 0.18 | 0.8333 |

Source: ERROR(YEAR*SEASON)

| DF | ANOVA SS | Mean Square |
|---|---|---|
| 16 | 18.80000000 | 1.17500000 |

```
Greenhouse-Geisser Epsilon = 0.7073
     Huynh-Feldt Epsilon = 0.9221


Level of     ------------READ1----------     ------------READ2----------
SES       N    Mean            SD              Mean            SD

HIGH      5  61.8000000    1.92353841      52.8000000    4.14728827
LOW       5  58.0000000    2.23606798      43.4000000    2.60768096

Level of     ------------READ3----------     ------------READ4----------
SES       N    Mean            SD              Mean            SD

HIGH      5  61.2000000    2.58843582      56.8000000    4.43846820
LOW       5  57.2000000    2.38746728      47.8000000    1.78885438

Level of     ------------READ5----------     ------------READ6----------
SES       N    Mean            SD              Mean            SD

HIGH      5  61.8000000    3.27108545      63.2000000    4.32434966
LOW       5  57.6000000    3.57770876      53.2000000    4.43846820
```

We will discuss the statistical results later, after two alternate programs have been presented. However, the program above which uses the REPEATED statement, produces two statistics, Greenhouse-Geisser-Epsilon and the Huynh-Feldt-Epsilon which we will explain here. There are some assumptions in repeated measures designs which are rather complicated. You may see these mentioned as symmetry tests or even sphericity tests. Somewhat simplified, what is being tested is the assumption that the variances and correlations are the same among the various dependent variables (See Edwards, 1985 mentioned in Chapter 1). It is more or less an extension of the assumption of equal variances in t-tests or ANOVA. SAS software provides tests for these assumptions listed as the Greenhouse-Geisser and the Huynh-Feldt tests. If they are nonsignificant, you may proceed with the analysis. If they are significant, under certain circumstances, you can still do the analysis, but you'll need some consultation with a statistician.

We will now present the other two programs that analyze this experiment without use of the REPEATED statement. Here is a second method: We have arranged our data so that each line represents one **cell** of our design. In practice, this would be tedious, but it will help you understand the last program for this problem in which all data for a subject are read on one line and the data set is transformed to look like this one.

```
DATA;
INPUT SUBJ SES $ YEAR SEASON $ READ;
CARDS;
1 HIGH 1 SPRING 61
1 HIGH 1 FALL 50
1 HIGH 2 SPRING 60
1 HIGH 2 FALL 55
1 HIGH 3 SPRING 59
1 HIGH 3 FALL 62
2 HIGH 1 SPRING 64
        etc.
```

This final approach places all the data for each subject on one line and does **not** use the REPEATED statement. This was the approach that had to be used before SAS software supported the REPEATED statement of ANOVA and GLM. We have left it in since the programming techniques are useful and are needed in cases where **all** the factors are repeated. As we have mentioned before, since there is an alternative easier method (above), you may skip the more elaborate program below and not sacrifice anything in the way of statistical understanding:

### ALTERNATIVE PROGRAM FOR READING EXPERIMENT

```
1    PROC FORMAT;
2        VALUE FSEASON 1=SPRING 2=FALL;
3        VALUE FSES 1=HIGH 2=LOW;
4    DATA;
5    DO SES=1 TO 2;
6        DO I=1 TO 5;
7            SUBJ+1;
8            DO YEAR=1 TO 3;
9                DO SEASON=1 TO 2;
10                   INPUT READ @;
11                   OUTPUT;
12                   END;
13               END;
14           END;
15       END;
16   FORMAT SEASON FSEASON. SES FSES.;
17   DROP I;
18   CARDS;
     61 50 60 55 59 62
     64 55 62 57 63 63
             etc.
```

(Note: The indentation is **not** necessary. It is used as a visual aid to keep the DO loops straight.)

This program is **not** as complicated as it may seem at first glance. Lines 1 through 4 create two formats to be used later. Our data will be arranged in the same order as they appear in the diagram of the experimental design. All high-SES students will be read followed by the low-SES students. Each student will have a spring/ fall set of reading comprehension scores for each of the three years. Since we have 5 subjects in each SES group, the loop in line 6 will go from 1 to 5; the next loop in line 8 initially sets the variable YEAR to 1 and the next loop in line 9 starts at SEASON=1. Finally, in line 10, a data value is read. The "@" causes the next input statement to read data from the same line of data. When line 11, OUTPUT, is executed, a line is written in the SAS data set consisting of the variables

SES SUBJ YEAR SEASON READ

Note that the variable I is not included because of the DROP statement in line 17. It is not necessary to drop I; we simply don't need it. We could leave it in the data set and just not use it. You should also be aware that DO loops can have character arguments. Thus, we could have written "DO SES='HIGH', 'LOW';" and "DO SEASON='SPRING', 'FALL';." Be careful, because if the length of the first value in the DO loop is shorter than the other levels, the program will truncate the length of the character variable to the first length it encounters. To avoid this problem, either pad the first value with blanks to be equal to the length of the longest value or use a LENGTH statement to define the length of the character variable. For example: "DO GRP = 'ONE ','TWO','THREE';"

**END OF ALTERNATIVE PROGRAM**

---

Now, regardless of the SAS data statements you used, the ANOVA statements will be the following:

```
PROC ANOVA;
   TITLE 'READING COMPREHENSION STUDY';
   CLASSES SUBJ SES YEAR SEASON;
   MODEL READ = SES SUBJ(SES)
              YEAR SES*YEAR YEAR*SUBJ(SES)
              SEASON SES*SEASON SEASON*SUBJ(SES)
              YEAR*SEASON SES*YEAR*SEASON YEAR*SEASON*SUBJ(SES);
   MEANS YEAR / DUNCAN E=YEAR*SUBJ(SES);
   MEANS SES SEASON SES*YEAR SES*SEASON YEAR*SEASON
       SES*YEAR*SEASON;
   TEST H=SES                    E=SUBJ(SES);
   TEST H=YEAR SES*YEAR          E=YEAR*SUBJ(SES);
   TEST H=SEASON SES*SEASON      E=SEASON*SUBJ(SES);
   TEST H=YEAR*SEASON SES*YEAR*SEASON
                                 E=YEAR*SEASON*SUBJ(SES);
```

As before, we specify hypotheses and error terms with TEST statements following our MODEL and we include the appropriate error terms with the Duncan requests.

Have our original ideas about "slippage" been confirmed by the data?

First, let us examine each of the main effects and their interactions:

```
SOURCE                    F VALUE              PR > F
-----------------------------------------------------------
SES                        13.54               .0062
YEAR                       26.91               .0001
SEASON                    224.82               .0001
SES*YEAR                     .11               .8962
SES*SEASON                 37.05               .0003
YEAR*SEASON               112.95               .0001
SES*YEAR*SEASON             0.18               .8333
```

Next are the means for each of the main effects and two-way interactions:

```
Level of          --------------READ--------------
SES        N        Mean                 SD

1          30      59.6000000          4.92425384
2          30      52.8666667          6.23523543

Level of          --------------READ--------------
SEASON     N        Mean                 SD

1          30      59.6000000          3.22276386
2          30      52.8666667          7.26224689

Level of  Level of        --------------READ--------------
SES       YEAR     N        Mean                 SD

1         1        10      57.3000000          5.63816361
1         2        10      59.0000000          4.13655788
1         3        10      62.5000000          3.68932394
2         1        10      50.7000000          8.02842174
2         2        10      52.5000000          5.33853913
2         3        10      55.4000000          4.45221543
```

| Level of SES | Level of SEASON | N | -------------READ------------- Mean | SD |
|---|---|---|---|---|
| 1 | 1 | 15 | 61.6000000 | 2.47270817 |
| 1 | 2 | 15 | 57.6000000 | 5.96178305 |
| 2 | 1 | 15 | 57.6000000 | 2.61315354 |
| 2 | 2 | 15 | 48.1333333 | 5.06904706 |

| Level of YEAR | Level of SEASON | N | -------------READ------------- Mean | SD |
|---|---|---|---|---|
| 1 | 1 | 10 | 59.9000000 | 2.80673792 |
| 1 | 2 | 10 | 48.1000000 | 5.93389510 |
| 2 | 1 | 10 | 59.2000000 | 3.15524255 |
| 2 | 2 | 10 | 52.3000000 | 5.71644800 |
| 3 | 1 | 10 | 59.7000000 | 3.91719855 |
| 3 | 2 | 10 | 58.2000000 | 6.69659947 |

What conclusions can we draw from these results?

(1) High-SES students have higher reading comprehension scores than low-SES students ($F=13.54$, $p=.0062$).

(2) Reading comprehension increases with each year ($F=26.91$, $p=.0001$). However, this increase is due partly to the smaller "slippage" in the later years (see 5 below).

(3) Students had higher reading comprehension scores in the spring compared to the following fall ($F=224.82$, $p=.0001$).

(4) The "slippage" was greater for the low-SES students (there was a significant SES*SEASON interaction $F=37.05$, $p=.0003$).

(5) "Slippage" decreases as the students get older (YEAR*SEASON is significant $F=112.95$, $p=.0001$).


Repeated measures designs can be a powerful ally of the applied researcher. They can also be a little bit tricky. For example, in our coffee study, even though we randomized  the order of first drinking the coffee  with dinner or breakfast,  there may be an effect we're overlooking.  It may be that one (or all) of the brands take a  little "getting used to."  This could result  in subjects  preferring  their second drinking of  the  coffee (whether  breakfast **or** dinner).  We are ignoring this in our study and  maybe  we shouldn't  be.  Had we **not** randomized which  drinking came first,  we  would have **con founded** drinking  order with MEAL.  The best way to make sure  that  you  are getting what  you  want  out  of  a repeated  measures  design  is to consult a  text  which deals solely with the design  and statistical issues involved. (Winer does an excellent job of this.)

# Chapter 8   Multiple Regression Analysis

## A. INTRODUCTION

Multiple regression analysis is a method for relating two or more independent variables to a dependent variable. In **most** applications, all of the variables are continuous variables (as opposed to categorical variables such as "sex" or "type of medication," where analysis of variance is the technique of choice). There are two rather distinct uses of multiple regression, and they will be addressed separately. The first use is in studies where the levels of the independent variables have been experimentally controlled (such as amount of medication and number of days between dosages). This use will be referred to as **"designed regression."** The second use involves settings where a sample of subjects have been observed on a number of naturally occurring variables (age, income, level of anxiety, etc.) which are then related to some outcome of interest. This use of regression will be referred to as **"nonexperimental regression."**

It is fairly easy to misuse regression. We will try to note some popular pitfalls, but we cannot list them all. A rule of thumb is to use your common sense. If the results of an analysis don't make any sense, get help. Ultimately, statistics is a tool employed to help us understand life. Although understanding life can be tricky, it is not usually perverse. Before accepting conclusions which seem silly based on statistical analyses, consult with a veteran data analyst. Most truly revolutionary results from data analyses are based on keypunching errors.

## B. DESIGNED REGRESSION

Imagine a researcher interested in the effects of scheduled exercise and the use of a stimulant on weight loss. She constructs an experiment on 24 college sophomores in which 4 levels of stimulant and 3 levels of exercise are used. There are 24 subjects in the experiment and each is randomly assigned to a level of exercise and stimulant such that two students are in each of the 12 (3X4) possible combinations of exercise and stimulant. After 3 weeks of participation, a measure of weight loss (pre/post weight) is obtained on each subject. The data for the experiment might look as shown below:

Data for Weight Loss Experiment

| SUBJECT | STIMULANT (mg/day) | EXERCISE (hrs/week) | WEIGHT LOSS (pounds) |
|---------|--------------------|---------------------|----------------------|
| 1 | 100 | 0 | -4 |
| 2 | 100 | 0 | 0 |
| 3 | 100 | 5 | -7 |
| 4 | 100 | 5 | -6 |
| 5 | 100 | 10 | -2 |
| 6 | 100 | 10 | -14 |
| 7 | 200 | 0 | -5 |
| 8 | 200 | 0 | -2 |
| 9 | 200 | 5 | -5 |
| 10 | 200 | 5 | -8 |
| 11 | 200 | 10 | -9 |
| 12 | 200 | 10 | -9 |
| 13 | 300 | 0 | 1 |
| 14 | 300 | 0 | 0 |
| 15 | 300 | 5 | -3 |
| 16 | 300 | 5 | -3 |
| 17 | 300 | 10 | -8 |
| 18 | 300 | 10 | -12 |
| 19 | 400 | 0 | -5 |
| 20 | 400 | 0 | -4 |
| 21 | 400 | 5 | -4 |
| 22 | 400 | 5 | -6 |
| 23 | 400 | 10 | -9 |
| 24 | 400 | 10 | -7 |

These data could be analyzed either as a 3X4 analysis of variance or as a two-variable multiple regression. The regression approach assumes that the effects of exercise and medication increase linearly (i.e., in a straight line); the ANOVA model makes no such assumption. If we use the multiple regression approach, the following program will provide the desired results:

```
1   DATA REGRESSN;
2   INPUT ID DOSAGE EXERCISE LOSS;
3   CARDS;
    1 100 0 -4
    2 100 0 0
    3 100 5 -7
      etc.
4   PROC REG;
5      MODEL LOSS = DOSAGE EXERCISE / P R;
```

Lines 1-3 define the data set. Line 4 specifies the PROC REGRESSION. Line 5 specifies the MODEL to be estimated. This model simply calls for the dependent variable LOSS to be regressed upon the independent variables DOSAGE and EXERCISE. The options "P" and "R" specify that we want predicted values and residuals to be computed. The output from this program is presented below:

DEP VARIABLE: LOSS

| SOURCE | DF | SUM OF SQUARES | MEAN SQUARE | F VALUE | PROB>F |
|---|---|---|---|---|---|
| MODEL | 2 | 162.971 | 81.485417 | 11.185 | 0.0005 |
| ERROR | 21 | 152.988 | 7.285119 | | |
| C TOTAL | 23 | 315.958 | | | |
| ROOT MSE | | 2.699096 | R-SQUARE | 0.5158 | |
| DEP MEAN | | -5.458333 | ADJ R-SQ | 0.4697 | |
| C.V. | | -49.4491 | | | |

| VARIABLE | DF | PARAMETER ESTIMATE | STANDARD ERROR | T FOR H0: PARAMETER=0 | PROB > \|T\| |
|---|---|---|---|---|---|
| INTERCEP | 1 | -2.562500 | 1.508841 | -1.698 | 0.1042 |
| DOSAGE | 1 | 0.001166667 | 0.004927852 | 0.237 | 0.8151 |
| EXERCISE | 1 | -0.637500 | 0.134955 | -4.724 | 0.0001 |

| ID | ACTUAL | PREDICT VALUE | STD ERR PREDICT | RESIDUAL | STD ERR RESIDUAL | STUDENT RESIDUAL |
|----|--------|---------------|-----------------|----------|------------------|------------------|
| 1 | -4.000 | -2.446 | 1.142 | -1.554 | 2.445 | -0.636 |
| 2 | 0 | -2.446 | 1.142 | 2.446 | 2.445 | 1.000 |
| 3 | -7.000 | -5.633 | 0.921917 | -1.367 | 2.537 | -0.539 |
| 4 | -6.000 | -5.633 | 0.921917 | -.366667 | 2.537 | -0.145 |
| 5 | -2.000 | -8.821 | 1.142 | 6.821 | 2.445 | 2.789 |
| 6 | -14.000 | -8.821 | 1.142 | -5.179 | 2.445 | -2.118 |
| 7 | -5.000 | -2.329 | 0.905304 | -2.671 | 2.543 | -1.050 |
| 8 | -2.000 | -2.329 | 0.905304 | 0.329167 | 2.543 | 0.129 |
| 9 | -5.000 | -5.517 | 0.603536 | 0.516667 | 2.631 | 0.196 |
| 10 | -8.000 | -5.517 | 0.603536 | -2.483 | 2.631 | -0.944 |
| 11 | -9.000 | -8.704 | 0.905304 | -.295833 | 2.543 | -0.116 |
| 12 | -9.000 | -8.704 | 0.905304 | -.295833 | 2.543 | -0.116 |
| 13 | 1.000 | -2.212 | 0.905304 | 3.212 | 2.543 | 1.263 |
| 14 | 0 | -2.212 | 0.905304 | 2.212 | 2.543 | 0.870 |
| 15 | -3.000 | -5.400 | 0.603536 | 2.400 | 2.631 | 0.912 |
| 16 | -3.000 | -5.400 | 0.603536 | 2.400 | 2.631 | 0.912 |
| 17 | -8.000 | -8.587 | 0.905304 | 0.587500 | 2.543 | 0.231 |
| 18 | -12.000 | -8.587 | 0.905304 | -3.413 | 2.543 | -1.342 |
| 19 | -5.000 | -2.096 | 1.142 | -2.904 | 2.445 | -1.188 |
| 20 | -4.000 | -2.096 | 1.142 | -1.904 | 2.445 | -0.779 |
| 21 | -4.000 | -5.283 | 0.921917 | 1.283 | 2.537 | 0.506 |
| 22 | -6.000 | -5.283 | 0.921917 | -.716667 | 2.537 | -0.283 |
| 23 | -9.000 | -8.471 | 1.142 | -.529167 | 2.445 | -0.216 |
| 24 | -7.000 | -8.471 | 1.142 | 1.471 | 2.445 | 0.601 |

SUM OF RESIDUALS -7.99361E-15    SUM OF SQUARED RESIDUALS 152.9875

Note: The output was truncated somewhat so that it would fit conveniently on the page.


The output begins with an analysis of variance table which looks much as it would from a standard ANOVA. We can see that there are 2 degrees of freedom in this model, one for EXERCISE and one for DOSAGE. There is only one degree of freedom for each since the regression estimates a single straight line for each variable rather than estimating a number of cell means. The sum of squares for the model (162.971) tells us how much of the variation in weight loss is attributable to EXERCISE and DOSAGE. The mean square for the model (81.485) is the sum of squares (162.971) divided by the degrees of freedom for the model (2). This mean square is then divided by the mean square error (7.285) to produce the F statistic for the regression (11.185). The p value for this is reported as .0005. "C TOTAL" means "corrected total" and indicates the total degrees of freedom (23) and sum of squares (315.958) in the dependent variable. The corrected total degrees of freedom is always one less than the total sample size since one degree of freedom is used to estimate the grand mean. The ROOT MSE (2.699) stands for the square root of the mean square error and represents in standard deviation units the variation in the system not attributable to EXERCISE or DOSAGE. DEP Mean (-5.458) is simply the mean of the dependent variable (LOSS). The R-SQUARE (.5158) is the square of the multiple correlation of EXERCISE and DOSAGE with LOSS. It is the proportion of variance in LOSS explained by (attributable to) the independent variables. ADJ R-SQ (.4697) is the adjusted R-square. The adjusted R-square takes into account how many variables were used in the equation and slightly lowers the estimate of explained variance. C.V. (-49.449) stands for coefficient of variation and is calculated by dividing the ROOT MSE by the mean and multiplying by 100. The C.V. is sometimes useful when the mean and standard deviation are related (such as in income data).

Having explained the terms in the analysis of variance table for the regression, let's summarize what meaning we can infer. Basically, the table indicates that the independent variables **were** related to the dependent variable (since the F was significant at p=.0005). Furthermore, we find that about 50% of the variation in weight loss is explained by the two experimental treatments. Many researchers are more interested in the R-square statistic than in the p-value since the R-square represents an estimate of how strongly related the variables were. The bottom half of the printout contains the estimates of the parameters of the regression equation. Three parameters are estimated: (1) the intercept, or constant term, (2) the coefficient for DOSAGE, and (3) the coefficient for EXERCISE. Each parameter estimate was based on 1 degree of freedom (always the case in regressions). For each parameter estimate, a standard error was estimated along with a t-statistic and a p-value for the t-statistic. The t-statistic is simply the parameter estimate divided by its standard error. The t-statistic is based on the number of degrees of freedom for the error term (21 in this example).

This half of the printout tells us that it was really EXERCISE that caused the weight loss. The regression coefficient for DOSAGE is not statistically significantly different from zero (p=.8151). The fact that the intercept was not significantly different from zero is irrelevant here. The intercept merely tells us where the regression line (or plane in this case) crosses the y-axis and does not explain any variation.

At this point, many researchers would run a new regression with DOSAGE eliminated, to refine the estimate of EXERCISE. Since this was a designed experiment, we would recommend leaving the regression as is for purposes of reporting. Dropping DOSAGE won't affect the estimated impact of EXERCISE since DOSAGE and EXERCISE are uncorrelated (by design). **When the independent variables in a regression are uncorrelated, the estimates of the regression coefficients are unchanged by adding or dropping independent variables.** When the independent variables **are** correlated, dropping and adding variables strongly affects the regression estimates.

**C. NONEXPERIMENTAL REGRESSION**

Many, if not most, regression analyses are conducted on data sets where the independent variables show some degree of correlation. These data sets, resulting from nonexperimental research, are common in all fields. Studies of factors affecting heart disease or the incidence of cancer, studies relating student characteristics to student achievement, and studies predicting economic trends all utilize nonexperimental data. The potential for a researcher to be misled by a nonexperimental data set is high; for a novice researcher, it is near certainty. We strongly urge consultation with a good text in this area (Pedhazur, **Multiple Regression in Behavioral Research** is excellent) or with a statistical consultant. Having presented this caveat,    let's  venture into  regression  analysis  for nonexperimental data sets.

## The Nature of the Data

There are usually many surface similarities between experimental and nonexperimental data sets. First, there are  one or more outcome or dependent variables. Second, there are several independent variables (sometimes quite a   few).   The  basic  difference  here  is  that  the independent variables are   correlated  with one another. This is because in nonexperimental studies one defines a population  of  interest  (people  who  have  had  heart attacks,  sixth grade students,  etc.),  draws a sample, and measures the variables of interest.  The goal of the study  is usually to explain variation in the  dependent variable by one or more of the independent variables. So far it sounds simple.

The problem is that the correlation among the inde- pendent  variables  causes the regression  estimates  to change  depending  on  which independent  variables  are being  used.  That is,  the impact of B on A depends  on whether C is in the equation or not.  With C omitted,  B can look very influential.  With C included,  the impact of B can disappear completely! The reason for this is as follows:   A  regression coefficient tells us the **unique** contribution  of an independent variable to a  dependent variable. That is, the coefficient for B tells us what B contributes all by itself with no overlap with any other variable.   If  B is the only variable in the  equation, this is no problem. But if we add C,  and if B and C are correlated,  then the **unique** contribution of B on A will be changed. Let's see how this works in an example.

The subjects are a random sample of 6th grade  stu- dents from Metropolitan City School District.  The  fol- lowing measures have been taken on the subjects:

(1)  ACH6 -  Reading achievement at the end of sixth grade.

(2) ACH5 - Reading achievement at the end of fifth grade.

(3) APT - A measure of verbal aptitude taken in the fifth grade.

(4) ATT - A measure of attitude toward school taken in fifth grade.

(5) INCOME - A measure of parental income (in thousands of dollars per year).

Our data set is listed below. (Note: These are **not** real data.)

| ID | ACH6 | ACH5 | APT | ATT | INCOME |
|----|------|------|-----|-----|--------|
| 1 | 7.5 | 6.6 | 104 | 60 | 67 |
| 2 | 6.9 | 6.0 | 116 | 58 | 29 |
| 3 | 7.2 | 6.0 | 130 | 63 | 36 |
| 4 | 6.8 | 5.9 | 110 | 74 | 84 |
| 5 | 6.7 | 6.1 | 114 | 55 | 33 |
| 6 | 6.6 | 6.3 | 108 | 52 | 21 |
| 7 | 7.1 | 5.2 | 103 | 48 | 19 |
| 8 | 6.5 | 4.4 | 92 | 42 | 30 |
| 9 | 7.2 | 4.9 | 136 | 57 | 32 |
| 10 | 6.2 | 5.1 | 105 | 49 | 23 |
| 11 | 6.5 | 4.6 | 98 | 54 | 57 |
| 12 | 5.8 | 4.3 | 91 | 56 | 29 |
| 13 | 6.7 | 4.8 | 100 | 49 | 30 |
| 14 | 5.5 | 4.2 | 98 | 43 | 36 |
| 15 | 5.3 | 4.3 | 101 | 52 | 31 |
| 16 | 4.7 | 4.4 | 84 | 41 | 33 |
| 17 | 4.9 | 3.9 | 96 | 50 | 20 |
| 18 | 4.8 | 4.1 | 99 | 52 | 34 |
| 19 | 4.7 | 3.8 | 106 | 47 | 30 |
| 20 | 4.6 | 3.6 | 89 | 58 | 27 |

The purpose of the study is to understand what underlies the reading achievement of the students in the district. The following program was written to analyze the data:

```
DATA NONEXP;
INPUT ACH6 ACH5 APT ATT INCOME;
CARDS;
7.5 6.6 104 60 67
6.9 6.0 116 58 29
    etc.
PROC STEPWISE;
   MODEL  ACH6  =  ACH5  APT ATT   INCOME /
                   FORWARD MAXR;
```

Although we can run a multiple regression as before (e.g., with PROC REG), we have specified that a stepwise regression is to be run with ACH6 as the dependent variable and ACH5, APT, ATT, and INCOME as independent variables. Furthermore, in the MODEL statement, we specified that a FORWARD selection technique should be used as well as a MAXR technique. Before examining the output, we should discuss briefly stepwise regression and nonexperimental data.

## D. STEPWISE REGRESSIONS

As mentioned earlier, with nonexperimental data sets, the independent variables are not truly "independent" in that they are usually correlated with one another. If these correlations are moderate to high (say 0.50 and above), then the regression coefficients are greatly affected by that particular subset of independent variables is in the regression equation. If there are a number of independent variables to consider, coming up with the best subset can be difficult. Stepwise regression (there are a family of different stepwise approaches) was developed to assist researchers in arriving at this optimal subset. Unfortunately, stepwise regression is frequently misused. The problem is that the solution from a purely statistical point of view is often not the best from a substantive perspective. That is, a lot of variance is explained but the regression doesn't make much sense. We'll discuss this more when we examine the printout. Stepwise regression examines a number of different regression equations. Basically, the goal of stepwise techniques is to take a set of independent variables and put them into a regression one at a time in a specified manner until all variables have been added or until a specified criterion has been met. The criterion is usually one of statistical significance such as: there are no more regressors that would be significant if entered or improvement in variance explained (the additional $R^2$ to be gained by entering the next best regressor) is too small to bother with.

SAS software allows for five different stepwise techniques:

(1) FORWARD--This starts with the best single regressor, then finds the best one to add to what exists, the next best, etc.

(2) BACKWARD--This starts will **all** variables in the equation, then it drops the worst one, then the next, etc.

(3) STEPWISE--This is similar to FORWARD except that there is an additional step in which all variables in each equation are checked again to see if they remain significant after the new variable has been entered.

(4) MAXR--This is a rather complicated procedure, but basically it tries to find the one-variable regression with the highest $R^2$, then the two-variable regression with the highest $R^2$, etc.

(5) MINR--This is very similar to the MAXR, except that the selection system is slightly different.

Now, let's examine the printout from the program:

```
FORWARD SELECTION PROCEDURE FOR DEPENDENT VARIABLE ACH6

STEP 1     VARIABLE ACH5 ENTERED          R SQUARE = 0.66909805
                                          C(P) =    1.87549647

                DF      SUM OF SQUARES   MEAN SQUARE       F     PROB>F

REGRESSION       1        12.17624633   12.17624633    36.40    0.0001
ERROR           18         6.02175367    0.33454187
TOTAL           19        18.19800000

                B VALUE    STD ERROR    TYPE II SS        F     PROB>F

INTERCEPT     1.83725236
ACH5          0.86756297   0.14380353   12.17624633    36.40    0.0001
-----------------------------------------------------------------------

STEP 2     VARIABLE APT ENTERED           R SQUARE = 0.70817380
                                          C(P) =    1.76460424

                DF      SUM OF SQUARES   MEAN SQUARE       F     PROB>F

REGRESSION       2        12.88734675    6.44367337    20.63    0.0001
ERROR           17         5.31065325    0.31239137
TOTAL           19        18.19800000

                B VALUE    STD ERROR    TYPE II SS        F     PROB>F

INTERCEPT     0.64269963
ACH5          0.72475202   0.16813652    5.80435251    18.58    0.0005
APT           0.01824901   0.01209548    0.71110042     2.28    0.1497
-----------------------------------------------------------------------

NO OTHER VARIABLES MET THE 0.5000 SIGNIFICANCE LEVEL FOR ENTRY
```

```
MAXIMUM R-SQUARE IMPROVEMENT FOR DEPENDENT VARIABLE ACH6

STEP 1     VARIABLE ACH5 ENTERED              R SQUARE = 0.66909805
                                              C(P) =     1.87549647

                    DF    SUM OF SQUARES  MEAN SQUARE       F    PROB>F

REGRESSION          1       12.17624633   12.17624633    36.40   0.0001
ERROR              18        6.02175367    0.33454187
TOTAL              19       18.19800000

                   B VALUE     STD ERROR    TYPE II SS       F    PROB>F

INTERCEPT       1.83725236
ACH5            0.86756297   0.14380353   12.17624633    36.40   0.0001
---------------------------------------------------------------------

THE ABOVE MODEL IS THE BEST  1 VARIABLE MODEL FOUND.


STEP 2     VARIABLE APT ENTERED              R SQUARE = 0.70817380
                                             C(P) =     1.76460424

                   DF    SUM OF SQUARES  MEAN SQUARE       F    PROB>F

REGRESSION          2       12.88734675    6.44367337    20.63   0.0001
ERROR              17        5.31065325    0.31239137
TOTAL              19       18.19800000

                   B VALUE     STD ERROR    TYPE II SS       F    PROB>F

INTERCEPT       0.64269963
ACH5            0.72475202   0.16813652    5.80435251    18.58   0.0005
APT             0.01824901   0.01209548    0.71110042     2.28   0.1497
---------------------------------------------------------------------

THE ABOVE MODEL IS THE BEST  2 VARIABLE MODEL FOUND.


STEP 3     VARIABLE ATT ENTERED              R SQUARE = 0.71086255
                                             C(P) =     3.61935632

                   DF    SUM OF SQUARES  MEAN SQUARE       F    PROB>F

REGRESSION          3       12.93627670    4.31209223    13.11   0.0001
ERROR              16        5.26172330    0.32885771
TOTAL              19       18.19800000

                   B VALUE     STD ERROR    TYPE II SS       F    PROB>F
```

```
INTERCEPT    0.80013762
ACH5         0.74739939    0.18222852    5.53198290    16.82    0.0008
APT          0.01972808    0.01298905    0.75861687     2.31    0.1483
ATT         -0.00797735    0.02068119    0.04892995     0.15    0.7048
```

THE ABOVE MODEL IS THE BEST  3 VARIABLE MODEL FOUND.

MAXIMUM R-SQUARE IMPROVEMENT FOR DEPENDENT VARIABLE ACH6

STEP 4     VARIABLE INCOME ENTERED             R SQUARE = 0.72232775
                                               C(P) =     5.00000000

|  | DF | SUM OF SQUARES | MEAN SQUARE | F | PROB>F |
|---|---|---|---|---|---|
| REGRESSION | 4 | 13.14492048 | 3.28623012 | 9.76 | 0.0004 |
| ERROR | 15 | 5.05307952 | 0.33687197 | | |
| TOTAL | 19 | 18.19800000 | | | |

|  | B VALUE | STD ERROR | TYPE II SS | F | PROB>F |
|---|---|---|---|---|---|
| INTERCEPT | 0.91164562 | | | | |
| ACH5 | 0.71373964 | 0.18932981 | 4.78747493 | 14.21 | 0.0019 |
| APT | 0.02393740 | 0.01419278 | 0.95826178 | 2.84 | 0.1124 |
| ATT | -0.02115577 | 0.02680560 | 0.20983199 | 0.62 | 0.4423 |
| INCOME | 0.00898581 | 0.01141792 | 0.20864378 | 0.62 | 0.4435 |

THE ABOVE MODEL IS THE BEST  4 VARIABLE MODEL FOUND.

Since a forward selection was requested first, that is what was run first. In step 1, the technique picked ACH5 as the first regressor since it had the highest correlation with the dependent variable ACH6. The RSQUARE (variance explained) is 0.669, which is quite high. "$C_p$" is a statistic used in determining how many variables to use in the regression. (It is too detailed to explain here; consult the SAS manual.) The remaining statistics are the same as for the PROC REG program run earlier. On step 2, the technique determined that adding APT would lead to the largest increase in $R^2$. We notice however, that RSQUARE has only moved from 0.669 to 0.708, a slight increase. Furthermore, the regression coefficient for APT is nonsignificant (p=.1497). This indicates that APT doesn't tell us much more than we already knew from ACH5. Most researchers would drop it from the model and use the one-variable (ACH5) model. After step 2 has been run, the forward technique indicates that no other variable would come close to being significant. In fact, no other variable would have a p value less than .50 (we usually require less than .05, although in regression analysis, it is not uncommon to set the inclusion level at .10).

The MAXR approach finds the best one-variable model, then the best two-variable model, etc. until the full model (all variables included) is estimated. As can be seen, with these data, both these techniques lead to the same conclusions: ACH5 is far and away the best predictor; it is a strong predictor; and no other variables would be included with the possible exception of APT.

There is a problem here, however. Any 6th grade teacher could tell you that the best predictor of 6th grade performance is 5th grade performance. But it doesn't really tell us very much else. It might be more helpful to look at APT, ATT, and INCOME **without** ACH5 in the regression. Also, it could be useful to make ACH5 the **dependent** variable and have APT, ATT, and INCOME be regressors. Of course, this is suggesting quite a bit in the way of regressions. There is another regression program which greatly facilitates looking at a large number of possibilities quickly. This PROC is called RSQUARE. The following lines will generate all of the regressions mentioned so far:

```
PROC RSQUARE;
   MODEL ACH6 = ACH5 APT ATT INCOME;
   MODEL ACH5 = APPT ATT INCOME;
```

The output from PROC RSQUARE looks like this:

```
N=     20      REGRESSION MODELS FOR DEPENDENT VARIABLE ACH6

NUMBER IN      R-SQUARE       VARIABLES IN MODEL
  MODEL

    1          0.10173375     INCOME
    1          0.18113085     ATT
    1          0.38921828     APT
    1          0.66909805     ACH5
-----------------------------------------
    2          0.18564520     ATT INCOME
    2          0.40687404     APT ATT
    2          0.45629702     APT INCOME
    2          0.66917572     ACH5 ATT
    2          0.66964641     ACH5 INCOME
    2          0.70817380     ACH5 APT
-----------------------------------------
    3          0.45925077     APT ATT INCOME
    3          0.66967022     ACH5 ATT INCOME
    3          0.71079726     ACH5 APT INCOME
    3          0.71086255     ACH5 APT ATT
-----------------------------------------
    4          0.72232775     ACH5 APT ATT INCOME
-----------------------------------------


N=     20      REGRESSION MODELS FOR DEPENDENT VARIABLE ACH5

NUMBER IN      R-SQUARE       VARIABLES IN MODEL
  MODEL

    1          0.13195687     INCOME
    1          0.26115761     ATT
    1          0.31693268     APT
-----------------------------------------
    2          0.26422291     ATT INCOME
    2          0.38784142     APT ATT
    2          0.41273318     APT INCOME
-----------------------------------------
    3          0.41908115     APT ATT INCOME
-----------------------------------------
```

The top part contains all of the RSQUARE's for every possible 1-, 2-, 3-, and 4-variable regression with ACH6 as the outcome variable. What you don't get in RSQUARE is any of the details. However, it **is** possible to glean a lot of information quickly from this table.

Let us say that you have just decided that you don't want ACH5 as a regressor. You can quickly see from the one-variable regressions that APT is the next best regressor ($R^2$=.389). The next question is,"What is the best two-variable regression?" and "Is the improvement large enough to be worthwhile?" Let's look at the two-variable regressions which have APT in them:

$R^2$ for      APT + ATT = .407
             APT + INCOME = .456

(Remember, we're eliminating ACH5 for now.)

APT and INCOME is best, and the gain is .067 (which is equal to 0.456 - 0.389). Is a 6.7% increase in variance explained worth including? Probably it is, although it may not be statistically significant with our small sample size. In explaining the regressions using ACH5 as an outcome variable, we can see that APT and INCOME look like the best bet there also. In interpreting these data, we might conclude that aptitude combined with parental wealth are strong explanatory variables in reading achievement. It is important to remember that statistical analyses must make substantive sense. The question arises here as to how these two variables work to influence reading scores. Some researchers would agree that APT is a psychological variable and INCOME is a sociological variable and the two shouldn't be mixed in a single regression. It's a bit beyond the scope of this book to speculate on this, but when running nonexperimental regressions, it is best to be guided by these two principles:

(1) Parsimony--less is more in terms of regressors. Another regressor will always explain a little bit more, but it often confuses our understanding of life.

(2) Common Sense--the regressors must bear a logical relationship to the dependent variable in addition to a statistical one. (ACH6 would be a great predictor of ACH5, but it is a logical impossibility.)

Finally, whenever regression analysis is used, the researcher should examine the simple correlations among the variables.

```
      PROC CORR;
         VAR APT ATT ACH5 ACH6;
```

will generate the following output:

| VARIABLE | N | MEAN | STD DEV | SUM | MINIMUM | MAXIMUM |
|---|---|---|---|---|---|---|
| ACH6 | 20 | 6.1100 | 0.97867 | 122.200 | 4.6000 | 7.5000 |
| ACH5 | 20 | 4.9250 | 0.92274 | 98.500 | 3.6000 | 6.6000 |
| APT | 20 | 104.0000 | 12.82678 | 2080.000 | 84.0000 | 136.0000 |
| ATT | 20 | 53.0000 | 7.74597 | 1060.000 | 41.0000 | 74.0000 |
| INCOME | 20 | 35.0500 | 16.18471 | 701.000 | 19.0000 | 84.0000 |

CORRELATION COEFFICIENTS / PROB > |R| UNDER H0:RHO=0 / N = 20

| | ACH6 | ACH5 | APT | ATT | INCOME |
|---|---|---|---|---|---|
| ACH6 | 1.00000 | 0.81798 | 0.62387 | 0.42559 | 0.31896 |
| | 0.0000 | 0.0001 | 0.0033 | 0.0614 | 0.1705 |
| ACH5 | 0.81798 | 1.00000 | 0.56297 | 0.51104 | 0.36326 |
| | 0.0001 | 0.0000 | 0.0098 | 0.0213 | 0.1154 |
| APT | 0.62387 | 0.56297 | 1.00000 | 0.49741 | 0.09811 |
| | 0.0033 | 0.0098 | 0.0000 | 0.0256 | 0.6807 |
| ATT | 0.42559 | 0.51104 | 0.49741 | 1.00000 | 0.62638 |
| | 0.0614 | 0.0213 | 0.0256 | 0.0000 | 0.0031 |
| INCOME | 0.31896 | 0.36326 | 0.09811 | 0.62638 | 1.00000 |
| | 0.1705 | 0.1154 | 0.6807 | 0.0031 | 0.0000 |

An examination of the simple correlations often leads to a better understanding of the more complex regression analyses. Here we can see why ATT, which shows a fairly good correlation with ACH6, was never included in a final model. It is highly related to INCOME

($r=.626$) and also to APT ($r=.497$). Whatever relationship it had with ACH6 was redundant with APT and INCOME. Also notice that INCOME and APT are unrelated, which contributes to their being included in the regressions. The simple correlations also protect against misinterpretation of "suppressor variables." These are a little too complex to discuss here. However, they can be spotted when a variable is not significantly correlated with the dependent variable but in the multiple regression has a significant regression coefficient (usually negative). You should get help (from Pedhazur or from another text or a consultant) with interpreting such a variable.

# Reference Section

This section is added for the second edition. It contains a chapter on SAS/PC®, a chapter on common questions and answers about SAS programs, and a chapter on advanced programming techniques. Many of the topics covered in this section have already been discussed elsewhere. However, we felt it would be convenient to place many of the tips and problem solutions in one place.

# Reference Chapter 1    SAC/PC

## A. INTRODUCTION

The arrival of SAS/PC® has made a dramatic change in the way SAS programmers have operated for the past 20 years or so. Users can now run SAS software on relatively inexpensive microcomputers. Data can be entered from data base management systems, word processors, spreadsheets, or directly into the SAS system. Output from SAS programs can be sent to printers or files, and incorporated in other documents. The power and flexibility of SAS/PC is enormous--but not without work on your part!

Unlike many "statistical packages" for microcomputers, SAS/PC is not "menu-driven." A novice user cannot simply choose items from a list and request statistics. A considerable effort must be made to learn a relatively complex system. The SAS Institute publishes thousands of pages of instructional manuals and offers video-taped courses and workshops. This book is an attempt to help users learn to use SAS software. Why work so hard when there are easier-to-use packages on the market? First, **POWER**. No package that we know of can come close to the data manipulation and statistical sophistication of SAS software. Also, once learned, programming with SAS software is faster than constantly choosing menu items and answering questions. SAS software puts its raw power right at your fingertips. Also, the Display Manager (to be discussed next) is cleverly designed to allow you to follow your program as it performs calculations or prints reports. In summation, SAS/PC is **not** for the casual, twice a year, user. It is geared for the researcher or professional who wants the power and flexibility of this extraordinary package.

To run SAS/PC, you need an IBM™ -compatible microcomputer with a "hard disk" and a minimum of 512K of memory. The SAS/PC family of products includes Base SAS® , SAS/STAT® , SAS/GRAPH® , and some specialized products such as SAS/IML™ (Interactive Matrix Language) SAS/AF™ (the Authoring Facility), and several others. Please note that SAS is a registered trademark

of SAS Institute Inc., Cary, NC. SAS/IML and SAS/AF are
trademarks of SAS Institute Inc., Cary, NC.

All users will need Base SAS (it uses almost 4
megabytes of your hard disk). In addition, to run most
of the statistical routines discussed in this book,
SAS/STAT (approximately 2.5 megabytes) will be needed.
Since this adds up to over 6 megabytes of storage, most
users will want at least a 20-megabyte hard disk. Our
experience with SAS/PC also leads us to recommend a full
640K of memory, especially for users with large data
sets or many variables. Finally, a coprocessor (either
an 8087 for an XT or an 80287 for an 80286-based
machine) is highly recommended. We have run SAS/PC on
IBM/XT's and AT's, Compaq 286, Leading Edge model D
(with a 20meg disk), Zenith Z200 (running PC-DOS), and
an AT&T 6300. It will most likely run on other com-
patibles but it would be a good idea to test it before
investing in a machine.

### B. USING A HARD DISK: SUBDIRECTORIES

If you are a new user of a "hard" disk or you have
been using one without using **subdirectories**, you should
read this section carefully. Since hard disks have the
capacity to hold large amounts of data (the smallest
size hard disk supplied with an IBM/XT is 10 million
characters), it is necessary to divide the disk into
logical subdivisions called **subdirectories**. Here's how
they work: When you boot the system (turn it on), you
will see a "C>" prompt. You are now in what is called
the **root** directory. If you issue the DOS "DIR" command,
you will see a list of all your files located in the
root directory. You may also see entries such as:

```
        SASDATA        <DIR>        10-15-85     3:46p
        SCOM           <DIR>         8-09-85    11:05a
        STEIN          <DIR>         6-04-86     1:43p
        UTIL           <DIR>         8-09-85    11:35a
        WS             <DIR>         1-04-80     8:59p
```

These are not files but subdirectories. Each sub-directory contains files. This allows us to group files logically. Without subdirectories, your list of files could extend over many screens. In addition, if you have more than 256 files in one directory, certain programs (such as BACKUP--DOS backup utility) will not work properly. Finally, SAS software is installed in a sub-directory and is usually run from a subdirectory.

To **create** a subdirectory, issue the command

MD\subdirectory name   (MD stands for **Make Directory**)

For example, to create a subdirectory call SASDATA, you would type

MD\SASDATA

To **C**hange to a sub**D**irectory, issue a **"CD\"** command. To enter the SASDATA subdirectory from the root direc-tory, you would type

CD\SASDATA

Once in a subdirectory, issuing a **DIR** command will display only those files located in the subdirectory. This is a convenient way to keep your files in logical subdivisions. Many people, the authors included, like to change the normal DOS prompt "C>" to one that shows in which subdirectory you are operating. This is done with a **PROMPT** command (one of the DOS commands). To change your prompt, issue the command: **PROMPT $P$G**. More con-veniently, place this command in your **AUTOEXEC.BAT** file. (See your DOS manual if you are unfamilar with this file.) Your prompt will now display the subdirectory name. When in the root directory, the prompt will look like this: **C:\>**. When in a subdirectory (SASDATA for example), it will look like this: **C:\SASDATA>**. By the way, to return to the root directory from a subdirec-tory, issue the command: **CD\**.

The only remaining subdirectory command is one to Remove Directories. It is: **RD\subdirectory name**. Before you issue this command, all files in the subdirectory must be removed. This is accomplished either by entering the command "DEL *.*" **from within the subdirectory** (careful, if you enter this command from the root directory or from another subdirectory, you will be deleting the wrong files!). An alternative would be to delete the files in the subdirectory while in the root directory. For example, to delete all files in the subdirectory SASDATA, you would type:

DEL \SASDATA

Once this is accomplished, you could proceed with removing the directory (RD). Note that the command DEL *.*, when issued from a subdirectory, deletes only those files in the subdirectory. One reason we like the prompt to display the subdirectory name is to make sure you know which subdirectory you are in, especially if you decide to globally delete files.

It is a good idea to be in a subdirectory before running a SAS program. Do not use the SAS subdirectory (i.e., the subdirectory where SAS programs are located) since installation of SAS software updates removes all programs from the SAS subdirectory. So if you named your subdirectory SASDATA, you would first change directories and then call invoke your SAS software:

**C:\>**CD\SASDATA
**C:\SASDATA>**SAS

Whenever you run SAS software, it creates two additional subdirectories of its own: **SASWORK** and **SASUSER**. If you are in a subdirectory, SASWORK and SASUSER will be subdirectories of your subdirectory. That is, you will have two new subdirectories that look like

SASDATA\SASWORK   and   SASDATA\SASUSER

This is important to know in case you decide to delete the SASDATA subdirectory. DOS insists that you remove all subdirectories underneath the one you want to delete. So, if you ran SAS programs from your SASDATA subdirectory, before removing it, you would first have to remove SASDATA\SASWORK and SASDATA\SASUSER.

## C. The SAS DISPLAY MANAGER

When you invoke the SAS system, your screen is divided into three "windows." On a color monitor, the upper window is pale blue, the middle window is gray, and the bottom window is dark blue. Each of these three windows has a separate function.

```
-OUTPUT----------------------------------------------------------
|Command ===>                                                    |
|                                                                |
|                                                                |
|                                                                |
|-LOG------------------------------------------------------------|
|Command ===>                                                    |
|                                                                |
|NOTE: Copyright ©   1985 SAS Institute Inc., Cary NC   27511, U.S |
|NOTE: SAS® Proprietary Software Release 6.xx Licensed to SAS    |
|                                                  Institute Inc |
|                                                                |
|-PROGRAM EDITOR-------------------------------------------------|
|Command ===>                                                    |
|                                                                |
| 00001                                                          |
| 00002                                                          |
| 00003                                                          |
|                                                                |
|----------------------------------------------------------------|
```

The bottom (dark blue) window is the **PROGRAM** window. You can enter your SAS program here or issue commands to bring in a SAS program from a file on your disk. This window contains an editor which allows you to edit and correct any errors in your program. There are also commands to save your program permanently on your disk. When complete, you can submit (run) your program with  a command or by pressing the appropriate   function key.

The  middle  window is the SAS LOG window.  The LOG window displays  the  program  lines  as the program ex-ecutes  and prints out messages about your data sets  as they  are  created  and  error messages  as  errors  are detected.  On a color monitor,  the messages about  your data  sets are in black,  the program lines are in blue, and the error messages are in red. SAS/PC makes such ef-fective  use of color  that we recommend a color monitor for frequent users of SAS software.  One of the   authors thought that color  was frivolous, but after running SAS programs on a color  monitor never returned to the world of monochrome again.

On top,  the  OUTPUT window displays the output  or results from your SAS program.  Again,  this will scroll by  as  the  SAS program is executing,  allowing you  to monitor the progress of the program.

We will now give  you  the details of each of these windows and the commands you will need to make effective use of  the  SAS  display manager.  Please refer to the SAS/PC manuals listed  in the Introduction to this  book for  the complete reference to the SAS/PC system.   This chapter is intended only as a quick reference to get you started. It is not complete and does not contain all the SAS/PC commands.

## D. THE PROGRAM WINDOW

When  you  first invoke the SAS system,  by  typing "SAS," the cursor will be located on the **Command** line of the **PROGRAM WINDOW**.  You can **zoom** (i.e., fill the screen with the current window,  making the other windows  tem-porarily disappear) by pressing the F7 key.  This key is a "toggle"--each time you press it, the display will al-ternate between the  three-window display and the "zoom" display.   To begin typing a program,  first "zoom"  the display by pressing the F7 function key.  Then press the RETURN key to move the cursor to the first numbered line of the PROGRAM WINDOW. Now you can type SAS statements, pressing the RETURN key at the end of each line. If you make a typing mistake, use the backspace key to erase the error. You may also use the arrow keys on the right side of the keyboard to position the cursor and use the "Del" key to edit your text. The SAS editor, by default, is in "overwrite" mode; that is, any text you type will overwrite existing text on that line. If you prefer an "insert" or "pushright" mode where any text you type pushes any existing text to the right, press the "Ins" key. You will then notice the "I" on the bottom right corner of the screen, indicating that you are in "Insert" mode. This key is also a "toggle"; press it again to change to "overwrite" mode.

As a test, try entering the following program:

```
00001 DATA TEST;
00002 INPUT X Y Z;
00003 CARDS;
00004 1 2 3
00005 4 5 6
00006 RUN;
00007 PROC PRINT;
00008 RUN;
```

When you have finished entering your program, you will want to **submit** it. To do this, press the F10 function key. If you can't remember which function key does what, try to remember F1 is the **HELP** key and **F2** shows the definition of all the other function keys.

Once you've submitted your program, the display will automatically "unzoom" to the three-window display. (An earlier version of SAS/PC did not do this, but all versions later than 6.02 will.) Next, you will see your program as it passes through the LOG window. This window will inform you that you have two observations and three variables. It will also display the time it took (in seconds) to process each step. Finally, your output will scroll by in the OUTPUT window. There are two ways to examine these two windows. One way is to place the cursor anywhere in a window you want to examine and then **zoom** the screen by pressing the "zoom" function key

(F7). This will cause the window that contains the cursor to fill the screen. You can now use the PgUp and PgDn keys on the numeric key pad to browse through the LOG or OUTPUT screens. A faster way to select a window is by using the function keys. The **F4** key will select the **OUTPUT** window, the **F3** key will select the **LOG** window, and the **F6** key will select the **PROGRAM** window. Again, remember that the **F7** key will zoom and unzoom the screen.

Suppose you made a mistake in entering the program. Below is the original program with an error introduced:

```
00001 DATA TEST;
00002 INPUT X Y Z;
00003 CARDS;
00004 1 2 3
00005 4 5 6
00006 RUN;
00007 PROC PRINY;
00008 RUN;
```

When you submit (F10) this program, an error message will appear in the SAS LOG (in red on a color display). At this point you will want to correct the program and run it again. You will probably want to look at the LOG window first to make sure you understand the error message. You can do this by pressing the F3 key to move to the LOG window and then F7 to zoom the screen. Then, using the PgUp and PgDn keys, you can read the messages in the SAS LOG. You next go back to the PROGRAM window (F6) to correct the program. Notice two things: One, the screen remains zoomed when you changed windows and, two, **our program has disappeared!** Don't panic--this is normal. As program statements are **submitted**, they are removed from the program window. To **recall** the portion of the program that was just submitted, press the **F9** (**recall**) key. Your program has returned. Now, using the PgUp, PgDn, and cursor arrow keys, locate the line of the program that contains the error (in our case, the PROC PRINT line, entered incorrectly as PROC PRINY). Correct the error by moving the cursor to the "Y" in PROC PRINY and replacing it with a "T". You can now submit the program again using the submit (F10) key.

To summarize the function keys, we have

| Action | Function Key |
|--------|--------------|
| Help | F1 |
| Key definition | F2 |
| LOG window | F3 |
| OUTPUT window | F4 |
| PROGRAM window | F6 |
| Zoom key | F7 |
| Recall program | F9 |
| Submit program | F10 |

Knowing how to **stop** a SAS program while it is executing is useful, especially if you have written a long program with many procedures and notice an **error** as the program executes. To stop a SAS program, enter a control C (i.e., hold down the control key while typing a "C"). The following message will appear:

```
==BREAK=====================================
| Press Y to cancel submitted statements, |
| T to halt data step/proc, N to continue.|
===========================================
```

Simply follow the directions to cancel all the following statements, to halt the currently executing data or proc step, or to continue.

A few more editing commands that you will need are described next. You may want to **insert** one or more lines between existing lines of your program. To do this, place the cursor anywhere within the **line number** of the line after which you want the new line or lines to go. To insert a single line, type an "i" and press the return key. A new line will be created and all lines below will be renumbered automatically. To insert more than one line, type an "i," a space, and the number of lines to insert. For example, to insert 4 lines, type "i 4" and press the return key. An insert function is illustrated below:

```
00001 DATA TEST;
00002 INPUT X Y Z;
00003 CARDS;
00004 1 2 3
00005 4 5 6              (Before insert)
00006 RUN;
00007 PROC PRINT;
00008 RUN;
```

```
00001 DATA TEST;
00002 INPUT X Y Z;
00003 CARDS;
00004 1 2 3
0i05 4 5 6        ("i" placed in number field))
00006 RUN;
00007 PROC PRINT;
00008 RUN;
```

```
00001 DATA TEST;
00002 INPUT X Y Z;
00003 CARDS;
00004 1 2 3
00005 4 5 6            (After insert)
00006 _                (Enter new line at 00006)
00007 RUN;
00008 PROC PRINT;
00009 RUN;
```

In a similar manner, single lines can be **deleted** with a "d" in the number field. To delete several lines, place "dd" in the first line to be deleted and move the cursor to the last line to be deleted and place a "dd" in the number field of that line. All lines from the first "dd" to and including the last "dd" will be deleted. A complete description of the SAS program editor can be found in the **SAS Language Guide** in the chapter on the Display Manager. (See list of references in the Introduction.)


## E. WORKING WITH FILES

One of the first file operations you will want to perform is saving your program for future use. To do this, first **recall** the program if it was already sub-

mitted.   Then,   move the cursor to the **command** line   by
pressing   the **HOME** key.   Now,   issue the **FILE** command as
follows:

    FILE 'filename'      where filename is a legal DOS
                         filename (may include drive and
                         path)

    Valid examples are:

    FILE 'XXX.SAS'                  FILE 'A:MYPROG'
    FILE '\SASDATA\TEST1.SAS'       FILE 'PROB1'

Another   use for the FILE command is to   write   SAS
**output**   to   a   file so that it may be printed   later   or
edited with your word processor.   To do this,   simply go
to the command line of the OUTPUT window (F4,   HOME) and
enter the FILE command,   just as you did in the   PROGRAM
window.   A special file name **'PRN'** will send your output
directly to your printer.

    FILE 'PRN'         (sends contents of current window
                       to your printer.)

Another important use of files is to read   external
data   in   an ASCII file.   This is a   common   requirement
where data were entered with a word processor or a database   management   program.   Suppose we created an   ASCII
file with our word processor and called it EXP1.DAT.   To
read   these data,   we will use an **INFILE** statement,   in
much the same way we read external   data   on a mainframe
computer.   As an example, imagine that the file EXP1.DAT
looks like this:

    001112053
    002218063
    003115568
      etc.

Our data layout is   ID col 1-3.   SEX col 4,   WEIGHT   col
5-7, and HEIGHT col 8-9.

Our program would then read

```
00001 DATA TEST;
00002 INFILE 'EXP1.DAT';
00003 INPUT ID 1-3 SEX 4 WEIGHT 5-7 HEIGHT 8-9;
00004 RUN;
00005 PROC ...
```

Notice that we precede the **INPUT** statement with an **INFILE** statement and that the CARDS statement is no longer needed.

Special care should be exercised when reading external data files. Unlike most mainframes which are "card" oriented, typical word processors do not **pad** each data line with blanks to make each line (record) of equal length. Therefore, even if your INPUT statement references columns, any record that is too short (e.g., has missing values on the end that were not entered as blanks) will cause the program **to go to the next record**. To avoid this, you can use the **missover** option on the INFILE statement. For example, if your file is called EXP1.DAT, your INFILE statement would read

INFILE 'EXP1.DAT' MISSOVER;

Note that some word processors create ASCII (American Standard Code for Information Interchange) files while others **do not**. Many of those that do not create ASCII files directly have utilities that will **convert** their format to ASCII. A special word to WordStar™ users. Use nondocument files only. In the normal document mode, Word Star uses special characters for such functions as page breaks. If a document is opened once and saved as a document file, you will have to convert the page break characters (carriage return followed by decimal 138) to a normal carrage return linefeed. Powerful word processors like PC-Write can do this.

Before we leave the subject of files, you may find it useful to write your SAS programs **outside** of the SAS Display Manager, using your favorite word processor. As long as you can produce an ASCII file with your word processor (either directly or through a conversion routine), you can take advantage of this feature. First, create the ASCII file containing the SAS program. Next, enter the SAS Display Manager. On the **command** line of the **program window,** type **INCLUDE** **'filename',** where filename is the name of your ASCII file. Note that the filename can include drive and path attributes. For example, suppose we just created a SAS program with our word processor and the file is called MYPROG and it is located in a subdirectory called MYSUB on the C drive. You would type

INCLUDE 'C:\MYSUB\MYPROG'

on the command line to bring the program into the **program window.** You can then submit the program (F10) or edit it further with the SAS editor.

## F. CREATING AND USING SAS DATA SETS

The more advanced user may want to create permanent SAS data sets. Normally, when a SAS data step executes, a temporary SAS data set is created. When you exit the SAS system (BYE), this data set disappears. If you are working on a project where the data set takes a long time to create and where you expect to run many procedures, a permanent SAS data set will save you time. The key to creating a permanent SAS data set is to use a **two-level** data set name and to add a **LIBNAME** statement to your program. A two-level data set name is of the form

LEVEL ONE.LEVEL TWO

where LEVEL ONE will indicate where the SAS data set is to be stored (in conjunction with a LIBNAME statement) and LEVEL TWO is the normal SAS data set name. The LIBNAME statement will "link" the LEVEL ONE name to a subdirectory on your disk. Suppose you want to save your

SAS data set in a subdirectory called BENZENE on your  C disk. The LIBNAME statement would be

    LIBNAME ANYNAME 'C:\BENZENE'

where  ANYNAME  is a name of your  choosing  which  must match the LEVEL ONE name on your data statement.  So, if the SAS data set name is EXPOSURE, you would have

    LIBNAME ANYNAME 'C:\BENZENE';
    DATA ANYNAME.EXPOSURE;
    INPUT ...

Notice  that  the LEVEL ONE name of  the  data  set matches the name on the LIBNAME statement. This program, when executed,  will  create  a SAS  data set called EX-POSURE  in  the  subdirectory  BENZENE.  SAS  software automatically adds the extension SSD to all SAS data set names,   so when you ask for a directory of your C disk, you will see an entry, EXPOSURE.SSD.  This is your  SAS data set. Once you have created a SAS data set, you will want  to use it.  Suppose you have just entered the  SAS Display Manager and want to run a PROC MEANS on the per-manent  SAS data set EXPOSURE.  Your program would  look like this:

    LIBNAME XXX 'C:\BENZENE';
    PROC MEANS DATA=XXX.EXPOSURE;
      VAR RBC WBC HEMO;

Notice  that the LEVEL ONE name is **not** the same  as when the data set was created.  All that is necessary is that  the LIBNAME and the LEVEL ONE name match (in  this case  we  used  XXX for the LIBNAME and  the  LEVEL  ONE name).

## G. DIFFERENCES BETWEEN MAINFRAME SAS AND SAS/PC

SAS/PC is  highly compatible with  mainframe  SAS software. One difference you  will find  is  with  **ARRAY** statements. Version 5  supports  the older form of ARRAY without  explicit subscripts while SAS/PC does not.  Any new  user  of SAS software should use the subscript form of the ARRAY statement exclusively so that there will be consistency between versions.

Another difference is that single quotes (') are required around titles, format values, labels, and character values of value statements. The following table illustrates these changes. Single quotes were always allowed (and sometimes required) for labels and format values. Now, you don't have to remember which special characters require quotes--you must use them all the time!

```
Old (Version 4) Style              New Style
---------------------------------------------------------
TITLE MY PROGRAM;                  TITLE 'MY PROGRAM';
VALUE S 1=WHITE 2=BLACK;           VALUE S 1='WHITE' 2='BLACK';
VALUE $S A=ONE B=TWO;              VALUE $S 'A'='ONE' 'B'='TWO';
LABEL X1=VARIABLE 1                LABEL X1='VARIABLE 1'
      X2=VARIABLE 2                      X2='VARIABLE 2';
```

# Reference Chapter 2    Questions and Answers

## A. INTRODUCTION

Having acted as consultants for many years, the authors have answered many questions about SAS statistical programming. This chapter attempts to answer some of the most common questions that students and professionals have asked about SAS software. Many of the topics covered in this chapter are covered in more detail in other sections of the book. More advanced topics will be found in the next chapter, "Advanced Programming Examples."

## B. WHAT IS THE DIFFERENCE BETWEEN A SAS DATA STEP AND A SAS PROCEDURE?

Whenever we see DATA statements in the middle of PROCs or PROC statements in the middle of DATA steps, we realize that we need to discuss the overall structure of SAS DATA and PROC steps.

The simplest form of a SAS program consists of a single DATA step followed by one or more PROCs. For example we could have

```
      -----     DATA SURVEY;
        |       INPUT ID AGE SEX QUES1-QUES5;
        |       CARDS;
   Data         001 23 1 1 3 2 4 3
   Step         002 34 2 2 2 1 2 1
        |       003 22 1 2 3 5 5 4
      -----     004 35 2 3 3 3 3 3
      -----     PROC MEANS;
        |          TITLE 'MY SURVEY';
   Proc            VAR AGE;
   Step         PROC FREQ;
        |          TABLES SEX QUES1-QUES5;
      ------
```

The DATA step is the place where the SAS data set is created. (See Chapter 1 for more details.) It starts with the word **DATA** and ends with the last line of data. It is here that we can make changes to our variables, create new variables, recode variables, or drop observations. Once the DATA step is completed, we can only make changes to the data set by copying it to another DATA set (using the SET statement). That is, after the DATA step, the first PROC, in essence, stops all DATA work. If we want to make more changes in the data set structure, we need to put in another DATA statement. You can't make data set changes in the middle of a PROC step.

Each PROCEDURE (PROC) that follows the data step will act on the most recently created data set unless we specify otherwise with a "DATA=" option following the procedure name.

Let's complicate matters a bit by adding a PROC **before** the first DATA step. We will use PROC FORMAT to create formats to be used in the data step. Here it is:

```
Proc          PROC FORMAT;
Step              VALUE SEX 1='MALE' 2='FEMALE';
-----         DATA SURVEY;
 |            INPUT ID AGE SEX QUES1-QUES5;
 |            FORMAT SEX SEX.;
 |            CARDS;
Data          001 23 1 1 3 2 4 3
Step          002 34 2 2 2 1 2 1
 |            003 22 1 2 3 5 5 4
-----         004 35 2 3 3 3 3 3
-----         PROC MEANS;
 |                TITLE 'MY SURVEY';
Proc              VAR AGE;
Step          PROC FREQ;
 |                TABLES SEX QUES1-QUES5;
------
```

This is one of the few times that a PROC step will precede the first DATA step. Suppose at this point we want to compute the average age for all respondents who answered either 1 or 2 to question 1. How do we do this? Some people might be tempted to do the following:

```
PROC MEANS;
    IF QUES1=1 OR QUES1=2;   WRONG WAY
    VAR AGE;
```

To make changes to the data set we will create a new data set like this:

(This section will be placed **below** the preceding program)

```
1.   DATA SUBSET;
2.   SET SURVEY;
3.   IF QUES1=1 OR QUES1=2;
4.   PROC MEANS;
5.      TITLE 'MEANS ON SUBSET DATA SET';
6.      VAR AGE;
```

The SET statement in line 2 will read each observation from the original data set "SURVEY" and the following line (3) will select only those observations that meet the designated criteria.

## C. HOW AND WHERE DO YOU CREATE NEW VARIABLES?

### Example 1--How to Create New Variables "Out of Thin Air."

Suppose we want to have our program generate a subject number for each person in a study. We can place an assignment statement in our data step like this:

```
DATA SURVEY;
INPUT AGE SEX $ HEIGHT WEIGHT;
SUBJ+1;
CARDS;
   data lines
```

The line "SUBJ+1" creates a new variable "out of thin air," called SUBJ which will have a value of 1 for the first observation, a value of 2 for the second observation, etc.

Another frequent reason to generate a new variable is to generate a **random number** for each subject. This might be needed, for example, to randomly assign subjects to a group or to test a statistical theory. Here is the code to do this:

```
DATA MYSTUDY;
INPUT ID X Y Z;
RANDOM = UNIFORM(0);
CARDS;
 data lines
```

Here, a new variable (RANDOM) will be included for each observation. The values of RANDOM will be a rational number from 0 to 1. Note that UNIFORM is a built-in SAS function. If you wanted a normally distributed random variable, the NORMAL function could be used instead. The argument of the UNIFORM or NORMAL function is called a **seed** and can be either zero, in which case the program uses the system time clock to compute the first random number, or you can use a 5-,6-, or 7-digit odd integer. Each time you run a SAS program where **you** supply the seed, SAS software will generate the **same series of random numbers.**

**Example 2--Creating New Variables from Old Ones**

The next program computes two new variables from existing variables in the data set. The first is average blood pressure, which is computed as the diastolic blood pressure plus one third the difference of the systolic and diastolic pressures. The formula is

```
AVEBP = DBP + (SBP - DBP)/3;
```

where AVEBP is the average blood pressure, SBP is the systolic blood pressure, and DBP is the diastolic blood pressure. The second variable to be computed is the natural logorithm of the heart rate. We have

```
DATA HEART;
INPUT ID HR SBP DBP;
AVEBP = DBP + (SBP - DBP)/3;
LOGHR = LOG(HR);
CARDS;
 data lines
PROC MEANS MAXDEC=2;
   VAR HR SBP DBP AVEBP LOGHR;
```

The new variables, AVEBP and LOGHR will be part of our data set and can be used in any procedure just like the variables that were listed in the INPUT statement.


## D. HOW DO YOU MODIFY VARIABLES?

As we mentioned above, the place to modify variables is in a DATA step. We will now show a few examples.

### Example 1--Transforming a Variable

For this example, we want to change the values of QUES5 so that they will be reversed. That is, we want to change 1 to 5, 2 to 4, 4 to 3, and 5 to 1. This is often done when questions are written as a **negative** question (i.e., the difference between "I often feel blue" and "I am happy most of the time"). The transformation we will use is simple:

```
                    QUES5=6-QUES5;
```

For any reverse transformation of this type, subtract the value of the variable from the number of response categories + 1.

Where do we place this statement? In the DATA step. We have

```
DATA SURVEY;
INPUT ID AGE SEX QUES1-QUES5;
QUES5=6-QUES5;
FORMAT SEX SEX.;
CARDS;
001 23 1 1 3 2 4 3
002 34 2 2 2 1 2 1
003 22 1 2 3 5 5 4
004 35 2 3 3 3 3 3
PROC MEANS;
    TITLE 'MY SURVEY';
    VAR AGE;
PROC FREQ;
    TABLES SEX QUES1-QUES5;
```

**Example 2--Recoding a Variable**

Here we want to create a new variable that will represent age groups. Again, the place to create this variable is in the data step. (See RECODING VARIABLES in Chapter 3 for an alternate method using FORMATS.)

```
DATA SURVEY;
INPUT ID AGE SEX QUES1-QUES5;
IF 0 LE AGE LE 25 THEN AGEGRP='0 TO 25 ';
   ELSE IF 26 LE AGE LE 50 THEN AGEGRP='26 TO 50';
   ELSE IF AGE GE 51 THEN AGEGRP='OVER 50 ';
FORMAT SEX SEX.;
CARDS;
```

We have chosen to create a **character** variable for AGEGRP so that we will save the trouble of creating a FORMAT. Notice that the **length** of the character variable will be defined **at the first true IF statement**. Therefore we have padded the values with blanks to be equal to the longest value. Alternatively, we could have used a **LENGTH** statement to assign a length initially. We are now free to use the variable AGEGRP in any following PROCs.

**E. HOW DO I SPECIFY A LIST OF VARIABLES?**

There are short-cut methods of specifying a list of variables, either in an INPUT statement, in a data step, or in a PROC. The first short-cut notation is the single dash (-). This allows us to refer to a list of variables that have **the same alphabetic stem** followed by a number. Thus, if we have variables ABC1, ABC2, and ABC3, we can write statements such as

```
INPUT ABC1-ABC3;
   .
   .
PROC MEANS;
   VAR ABC1-ABC3;
   .
   .
NEWVAR = MEAN (OF ABC1-ABC3);
```

The second method of specifying a list of variables is the double dash notation (--). This notation is used to specify all variables from the first to the last **in the order they are in the SAS data set.** So, if our INPUT statement is

▌     INPUT X Y Z ABC1-ABC3 AGE HEIGHT WEIGHT;

we can refer to all of our variables using the notation "X -- WEIGHT" or the three variables AGE HEIGHT and WEIGHT as "AGE -- WEIGHT." (Note that all numeric variables can be referred to as _NUMERIC_ in most places that allow variable lists.) Great caution should be exercised when using -- since the order of variables in a SAS data set is not always clear. Statements such as LENGTH and ARRAY can change the order. If in doubt, do not use the --, or run PROC CONTENTS to determine the order of the variables in the data set. (If you are using SAS/PC, use the VAR command to get a list of variables in order.)

**F. CAN I SPECIFY A LIST OF VARIABLES ON THE INPUT LINE WHEN I HAVE COLUMN DATA (WITHOUT SPACES)?**

Many people come to us with INPUT statements that look like this:

▌     INPUT ID 1-3 X1 4 X2 5 X3 6 X4 7 X5 8 X6 9;

This can be written more compactly with a **format list** like this:

▌     INPUT ID 1-3 (X1-X6) (1.);

This notation specifies that all the variables in the list (X1 through X6) are to be read with the 1. format (1 column of numeric data). What if the first X variable was not right next to the ID like this:

```
INPUT ID 1-3 X1 10 X2 11 X3 12 X4 13 X5 14 X6 15;  ▮
```

We can still use a format list but we must also use a **pointer** to indicate the column where X1 is located. The INPUT statement with pointer and format list looks like this:

```
INPUT ID 1-3 @10 (X1-X6) (1.);                    ▮
```

## G. HOW DO PROCEDURES OUTPUT DATA SETS?

Just when you thought you understood how all this works, we are now going to show you how to output a **DATA SET** with a **PROCEDURE**. The following example will use PROC MEANS to create an output data set.

### Computing a Mean for Each Subject

Suppose we are conducting a study where subjects are assigned to 1 of 5 treatment groups (GROUP) and each subject is measured either 2 OR 3 times. We are not interested in the individual measurements, but in the **mean** for each subject. Our original data set looks like this:

```
SUBJ      GROUP     TRIAL     SCORE     X     Y
-------------------------------------------------
  1         A         1         56       5     8
  1         A         2         59       5     7
  1         A         3         57       6     6
  2         A         1         62       9     8
  2         A         2         64       8     7
  3         B         1         77       4     4
  3         B         2         77       5     4
  3         B         3         79       6     6
               etc.
```

Notice that each measurement is a separate observation. Thus, we cannot use the **MEAN FUNCTION** to compute a subject mean. We can use either PROC MEANS or PROC SUMMARY to output a data set that contains a mean for each subject.

```
PROC SORT;
   BY SUBJ GROUP;
PROC MEANS NOPRINT;
   BY SUBJ GROUP;
   VAR SCORE X Y;
   OUTPUT OUT=MOUT MEAN=MSCORE MX MY;
```

Here is an example of a **PROC** that creates a **DATA** set. Notice that we have included a **BY** statement so that our output data set will contain a mean score for each subject and GROUP. Notice also that the data set must be sorted by the BY variable (unless it is already in sorted order). For large data sets, this is costly and PROC SUMMARY sould be used instead. The syntax of the output statement of PROC MEANS is

OUTPUT OUT=output data set name

MEAN=list of variables for the means in the same order as found on the var statement.

STD=If you want standard deviations include this keyword followed by another list of variable names (e.g., SDSCORE SDX, SDY).

ANY OTHER KEYWORDS USED WITH PROC MEANS= list of variable names.;

The data set MOUT will include the variables MSCORE, MX, and MY as well as all variables in the BY statement. The first few observations of the data set MOUT are

| OBS | SUBJ | GROUP | _TYPE_ | _FREQ_ | MSCORE | MX | MY |
|-----|------|-------|--------|--------|---------|---------|---------|
| 1 | 1 | A | 0 | 3 | 57.3333 | 5.33333 | 7.00000 |
| 2 | 2 | A | 0 | 2 | 63.0000 | 8.50000 | 7.50000 |
| 3 | 3 | B | 0 | 3 | 77.6667 | 5.00000 | 4.66667 |

The variables in the newly created data set (MOUT) are self-explanatory except for _TYPE_ and _FREQ_. The variable _FREQ_ tells us how many numbers were used to create the mean scores (in this case, the number of measurements per subject). The _TYPE_ variable is useful

only if a CLASS statement was used. Previous to version
6.0,  CLASS statements were used with PROC SUMMARY  and
not with PROC MEANS. The addition of a CLASS  statement
with version 6.0 makes PROC SUMMARY unnecessary. Using
the NOPRINT option with  PROC MEANS duplicates the functions formally found in PROC SUMMARY. Briefly,  a CLASS
variable allows us to compute  means  for every combination of  CLASS variables **without having to sort** the data
set first. This will save time and money.  If all  you
want  is  **cell means** then use the option **NWAY** of  PROC
MEANS or PROC SUMMARY. If you want means for  each combination  of CLASS variables,  do not include  the  NWAY
option. Below  are the PROC SUMMARY statements and  the
output  data  set from PROC SUMMARY (identical  to  PROC
MEANS):

```
PROC SUMMARY DATA=TEST NWAY;
    CLASS GROUP SUBJ;
    VAR SCORE X Y;
    OUTPUT OUT=SUMOUT MEAN=;
```

| OBS | GROUP | SUBJ | _TYPE_ | _FREQ_ | SCORE | X | Y |
|-----|-------|------|--------|--------|---------|---------|---------|
| 1 | A | 1 | 3 | 3 | 57.3333 | 5.33333 | 7.00000 |
| 2 | A | 2 | 3 | 2 | 63.0000 | 8.50000 | 7.50000 |
| 3 | B | 3 | 3 | 3 | 77.6667 | 5.00000 | 4.66667 |

Notice  that only the cell means (each  combination
of SUBJ and GROUP) are contained in the data set because
of the NWAY option. Finally,  an OUTPUT data set  from
PROC  SUMMARY (or PROC  MEANS)  without using the  NWAY
option is shown. Notice the mean for each combination of
CLASS variables is  produced.  Type 0 is the **grand mean**,
type 1 is the mean for each SUBJ, type 2 is the mean for
each GROUP,  and type 3 is the mean for each combination
of SUBJ and GROUP (cell means).

| OBS | GROUP | SUBJ | _TYPE_ | _FREQ_ | SCORE | X | Y |
|-----|-------|------|--------|--------|-------|---|---|
| 1 |   | . | 0 | 8 | 66.3750 | 6.00000 | 6.25000 |
| 2 |   | 1 | 1 | 3 | 57.3333 | 5.33333 | 7.00000 |
| 3 |   | 2 | 1 | 2 | 63.0000 | 8.50000 | 7.50000 |
| 4 |   | 3 | 1 | 3 | 77.6667 | 5.00000 | 4.66667 |
| 5 | A | . | 2 | 5 | 59.6000 | 6.60000 | 7.20000 |
| 6 | B | . | 2 | 3 | 77.6667 | 5.00000 | 4.66667 |
| 7 | A | 1 | 3 | 3 | 57.3333 | 5.33333 | 7.00000 |
| 8 | A | 2 | 3 | 2 | 63.0000 | 8.50000 | 7.50000 |
| 9 | B | 3 | 3 | 3 | 77.6667 | 5.00000 | 4.66667 |

## H. HOW DO I USE ARRAYS?

Beginning with version 5, SAS software supports explicit subscripts of arrays. Don't worry if that means nothing to you. If you follow the examples below, you will see the tremendous power of arrays. We have changed all the examples in this section to conform to the new, explicit subscript systax of versions 5 and 6. Even though version 5 will accept **both** forms of the ARRAY statement, we advise using only the newer form presented here. This enables you to transfer your programs between a mainframe and microcomputer. Furthermore, the older form may not be supported in future versions of SAS software.

Arrays in SAS programs allow us to execute SAS statements that will process a list of variables.

We will attempt to demonstrate the use of ARRAYS with a series of examples. Our first example shows how to change all values of 9 to "missing" for a series of variables.

**WITHOUT ARRAYS**

```
DATA;
INPUT A B C X1-X10;
IF A = 9 THEN A = .;
IF B = 9 THEN B = .;
IF C = 9 THEN C = .;
IF X1 = 9 THEN X1 = .;
IF X2 = 9 THEN X2 = .;
        etc.
CARDS;
```

**WITH ARRAYS**

```
DATA;
ARRAY M{*} A B C X1-X10;
INPUT A B C X1-X10;
DO I = 1 to 13;
    IF M{I} = 9 THEN M{I} = .;
    END;
CARDS;
```

In this example, the array name (M) represents a list of variables (A,B,C, X1-X10). Our "DO I = 1 to 13" statement "loops" through each variable in the ARRAY, so that M{I} when I = 1 stands for the variable A. When I = 2, M{I} represents B and so forth. All the statements between the "DO" and "END" will be executed as many times as indicated in the "DO" statement. The ARRAY statement is of the form

ARRAY{number of elements} Variable List

In our case, we used an asterisk "*" instead of the number of elements in the array (13). This is a shortcut allowed by SAS software. An asterisk says that the program should count the elements for us.

Our next example will use two arrays. One is a set of existing variables; the other, a set of variables to be computed.

**WITHOUT ARRAYS**               **WITH ARRAYS**

```
DATA;                           DATA;
INPUT A B HT1-HT10;             ARRAY HT{10} HT1-HT10;
HTCM1 = 2.54*HT1;               ARRAY HTCM{10} HTCM1-HTCM10;
HTCM2 = 2.54*HT2;               INPUT A B HT1-HT10;
    etc.                        DO J = 1 to 10;
CARDS;                              HTCM{J} = 2.54*HT{J};
                                    END;
                                CARDS;
```

Please note that the array names HT and HTCM are arbitrary and were chosen to be convenient for remembering what they represent.

Arrays can be especially useful when restructuring SAS data sets. Here is one of the examples from Chapter 3 and the same program using ARRAYS:

```
WITHOUT ARRAYS                      WITH ARRAYS

DATA DIAG2;                         DATA DIAG2;
SET DIAG1;                          SET DIAG1;
DX = DX1;                           ARRAY D{*} DX1-DX3;
IF DX NE . THEN OUTPUT;             DO I=1 TO 3;
DX = DX2;                              DX = D{I};
IF DX NE . THEN OUTPUT;                IF D{I} NE . THEN OUTPUT;
DX = DX3;                           END;
IF DX NE . THEN OUTPUT;             KEEP ID DX;
KEEP ID DX;
```

## I. WHAT IS THE DIFFERENCE BETWEEN THE FOUR TYPES OF SUMS OF SQUARES FROM ANOVA AND GLM?

SAS software offers us four different types of sums of squares (Type I - Type IV) for the user to choose among. Typically, a printout will include only type I and type III. Since type II and type IV are rather specialized, we refer the reader to the SAS manual for those. Since SAS views multiple regression and ANOVA as essentially the same procedure, it treats them similarly in terms of the printout. That is, the sum of squares part of the table looks the same whether you have run a regression or an ANOVA.

Basically, the difference between type I and type IV sums of squares comes into play when the cell sizes are not equal in ANOVA, or when the regressors are correlated in regression. (With equal cell sizes or uncorrelated regressors, type I and type III are identical.) When the cell sizes are unequal an ANOVA, in most cases, some correlation in the independent variables will result. In a regression where the regressors have not been assigned levels through experimental manipulation, the regressors will have some level of correlation. This correlation can become quite high (e.g., father's level of education as one regressor and mother's level of education as a second).

When independent variables (or regressors) are correlated, their relationship to the dependent (outcome) variable can overlap (again, consider the effect of mother's and father's levels of education on the achievement of children). When this overlap occurs, a question arises as to which variable should "get" the overlap. There are two models that are typically used. In one model (type I), the first effect listed on the model statement gets all the variation in the dependent variable that it can account for. (We use "effect" here instead of variables since interaction terms can be included.) Then the second effect gets all that it can account for **except** that part that overlapped with the first effect (it went with the first effect). Then the third effect gets all that it can (except for what went with the first two effects), etc. This is a form of a general class of procedures known as "stepwise."

The second type of approach (type III) only "gives" to each effect its **unique** contribution. That is, for any effect, the program first removes any overlap with **any** other effect, then examines the relationship between the effect in question and the outcome variable. This provides what is called the "unique" contribution of each effect.

So which effect to use? Generally speaking, type III. There are certain situations in which you know that you want to test effect A before effect B, but you'll usually want to examine the unique contribution of each effect. Two final points: One, if your cell sizes are nearly equal, the two models are quite similar. Two, on occasion, you can get a more significant finding under type III than type I for a given variable. This is due to a phenomenon called suppression, which is discussed in Pedhazur (See Chapter 1 for references).

## J. HOW DO I INTERPRET ERROR MESSAGES?

This section deals with some of the most common error messages generated by SAS programs and how to find and correct the error. The display of error messages is slightly different between version 5 and version 6.

1. **Variable not found.** This error message can result from a variety of errors. Most common is the **misspelling** of a variable name, either on the INPUT statement or elsewhere so that the names do not agree between the data step and a procedure that uses them. An example is shown below:

```
4      *COMMON ERRORS;
5      DATA ONE;
6      INPUT X Y Z;
7      CARDS;
8      PROC MEANS;
NOTE: The data set WORK.ONE has 2 observations and 3 variables.
NOTE: The DATA statement used 8.00 seconds.
9          VAR X U Z;
ERROR: Variable U not found.
```

2. **Syntax error detected.** This is another very common error caused either by mispelling a keyword or statement, or by using a statement that is not valid for a particular procedure. Below is the result of misspelling a keyword in a data step:

```
11     DATA TWO;
12     INTPU A B C;
INTPU A B C;
  -
ERROR: Syntax error detected

NOTE: Expecting one of the following :

; NAME NAME.NAME + : = ABORT ARRAY ATTRIB BY CALL CARDS CARDS4 DELETE
DISPLAY DO DROP FILE FORMAT GO GOTO IF INFILE INFORMAT INPUT KEEP LABEL
LABLE LENGTH LINK LIST LOSTCARD MERGE MISSING OUTPUT PUT RENAME RETAIN
RETURN RUN SELECT SET STOP SUBSTR UPDATE WINDOW [ {
```

3. **Invalid data error.** The most frequent problem here is a character value in a numeric field or an invalid number (such as 6.4.77). The SAS LOG will display the offending error line and substitute a **missing value** for the invalid data. It will then proceed with the following procedures:

```
17    DATA THREE;
18    INPUT SEX HEIGHT WEIGHT;
19    CARDS;
NOTE: Invalid data for SEX in column 1 :
RULE: ----+----1----+----2----+----3----+----4----+----5----+----6
   1> M 63 166
```

4. **Semicolon missing.** This is probably the **most common SAS programming error.** The error messages resulting from a missing semicolon can be very strange indeed. Look at the example below:

```
23    DATA FOUR;
24    INPUT SBP DBP HR
25    CARDS;
26    130 80 55
130 80 55
---
ERROR: Syntax error detected
```

Notice that the error message refers to the first **data** line! The reason for this is that the missing semicolon on the INPUT line causes the word CARDS to be read as a **variable name.** Then, when the first data line is encountered, the SAS compiler is looking for a SAS statement since it did not recognize the CARDS statement. The only advice we can give you is, "If the error message doesn't seem to make sense, look for a missing semicolon in the previous line or two."

## K. HOW CAN I ADD NEW OBSERVATIONS TO MY DATA?

This is a common question. Fortunately it has an easy answer. First, the easy method is to add the new observations as **raw data** and **recreate** the SAS data set. If this is impractical because the original data set is large, you can use a **SET** statement to add the new data. Here is an example where we will add some new, raw data to an existing SAS data set (EXIST):

```
DATA NEWDATA;
INPUT X Y Z;
CARDS;
1 2 3
4 5 6
DATA COMBINED;
SET MYLIB.EXIST NEWDATA;
```

Note that a DD statement (for batch OS systems), a filedef (for CMS or TSO), or a LIBNAME statement (SAS/PC) must be included to indicate to the computer where to find the SAS data set EXIST. Also, the data set NEWDATA must contain the **same variables** as the existing data set.

## L. HOW DO I ADD NEW VARIABLES TO MY DATA SET?

This is a more difficult problem than adding observations to a data set (above). If the data set is small, you may want to consider adding the new variables to the raw data and recreating the data set. Otherwise, the best way to proceed is to create a new SAS data set containing the new variables and an ID variable that has the same value for each observation in the original data set. For example, suppose the original data set contains SS(social security number), AGE, SEX, HEIGHT, and WEIGHT. We now want to add heart rate (HR) and blood pressure (BP) to the data set. We can create a new data set containing SS, HR, and BP. Then, using the MERGE statement, we can add the new variables to the data set. Here is an example:

```
DATA ORIG;
INPUT SS AGE SEX $ HEIGHT WEIGHT;
CARDS;
 data lines
DATA NEW;
INPUT SS HR BP;
CARDS;
 data lines
PROC SORT DATA=ORIG;
   BY SS;
PROC SORT DATA=NEW;
   BY SS;
DATA COMBINED;
   MERGE ORIG NEW;
   BY SS;
```

A very important part of this program is the BY statement following the MERGE. This assures us that the new variables from the data set NEW will be added to the proper observations of our original data set. We do **not** recommend using a MERGE **without** a **BY** variable except in very special cases. The program above will produce a data set that has all the observations from both ORIG and NEW, supplying missing values where no data exist. Suppose you wanted only observations for which new data values were collected. The **IN** option of the MERGE state-ment will do the trick. If we write

```
DATA COMBINED;
MERGE ORIG NEW (IN=XXX);
BY SS;
IF XXX;
```

we will accomplish this goal. The variable following IN= will be set to TRUE if an observation is contributed from the data set (in this case NEW). If we wanted only observations that had contributions from **both** data sets we would write

```
DATA COMBINED;
MERGE ORIG (IN=INORIG) NEW (IN=INNEW);
BY SS;
IF INORIG AND INNEW;
```

## M. HOW DO I COMBINE TWO OR MORE OBSERVATIONS INTO ONE?

In the section on longitudinal data analysis in Chapter 3 and in the sections on restructuring SAS data sets in Chapter 7, we showed you how to create several observations from a single observation, using the OUTPUT statement. There are times when the reverse process is necessary--creating one observation from several. Here is an example in which a researcher had a data set of the following form:

```
SUBJECT    TREAT    TIME     ZINC
  001        A       1       .33
  001        A       2       .44
  002        A       1       .67
  002        A       2       .61
  003        B       1       .55
  003        B       2       .78
  004        B       1       .35
  004        B       2       .88
                   etc.
```

This researcher wanted to run a two-way repeated measures analysis of variance (TREAT X TIME) where TIME was a repeated measure. Because of memory problems, the REPEATED option of PROC GLM had to be used. (Don't worry about the details of the analysis--this example is really to show you how to restructure a SAS data set.) Therefore, this researcher wanted a data set of the form

```
SUBJECT    TREAT    ZINC1    ZINC2
  001        A       .33      .44
  002        A       .67      .61
  003        B       .55      .78
  004        B       .35      .88
```

Notice that ZINC1 is the value of ZINC in the original data set at TIME=1 and ZINC2 is the value of ZINC at TIME=2. How do we accomplish this transformation? We will want to read the **first** observation for each subject and set the value of ZINC equal to ZINC1. Next, we will read the **last** observation for each subject and set the value of ZINC equal to ZINC2. The way to

recognize the **first** or **last** observation for each subject
is with the SAS internal FIRST. and LAST. variables (see
Section F in Chapter 3). In addition, we must remember
to **RETAIN** the value from the first observation so that
the SAS program will not set the value of ZINC1 to missing each time a new observation is read. Here is a
program to accomplish the transformation:

```
1       PROC SORT;
2          BY SUBJECT;
3       DATA TRANSFRM;
4       SET ORIG;   *WHERE ORIG IS THE ORIGINAL DATA SET;
5       BY SUBJECT;
6       RETAIN ZINC1;
7       IF FIRST.SUBJECT THEN ZINC1=ZINC;
8       IF LAST.SUBJECT THEN DO;
9          ZINC2=ZINC;
10         OUTPUT;
11         END;
12      KEEP SUBJECT TREAT ZINC1 ZINC2;
```

We remembered to sort the data set by subject
before we used the BY statement in line 5. We also included a RETAIN statement (line 6) so the value of ZINC1
would not be set to missing when the last observation
for our subject was read. With a little practice,
restructuring SAS data sets will become routine. Even
for the experienced programmer, however, we recommend
that you use a PROC PRINT to check that the data set is
really the one you want.

# Reference Chapter 3   Selected Programming Examples

This chapter contains a number of common applications. It serves two functions: one is to allow you to use any of the programs here, with modification, if you have a similar application; the other is to demonstrate SAS programming techniques.

## A. TEST SCORING, ITEM ANALYSIS, AND TEST RELIABILITY

This first example solves a common problem (scoring a test) as well as using many SAS features that cause difficulty for SAS users. First, we will write a program to score a test. The first line of data will be an **answer key** followed by student responses. The method here is to read the answer key differently from the student responses.

```
1      *PROGRAM TO SCORE A TEST;
2      DATA SCORE;
3      RETAIN KEY1-KEY10;
4      ARRAY KEY{*} KEY1-KEY10;
5      ARRAY ANS{*} ANS1-ANS10; *ANS ARE STUDENT ANS;
6      ARRAY S{*} S1-S10; *THE S'S ARE SCORED RESPONSES;
7      IF _N_ = 1 THEN DO;
8         INPUT @11 (KEY1-KEY10)(1.);
9         DELETE;
10        END;
11     ELSE DO;
12        INPUT SS 1-10 (ANS1-ANS10)(1.);
13        DO I = 1 TO 10;
14           IF KEY{I} = ANS{I} THEN S{I}=1;
15           ELSE S{I}=0;
16           END;
17        RAW=SUM (OF S1-S10);
18        PERCENT=100*RAW/10;
19        DROP I KEY1-KEY10;
20        END;
21     FORMAT SS SSN11.;
22     CARDS;
       ANS KEY    1234554321
       102231324 1234323321
       101202333 3334254321
       102010222 1233544211
               etc.
23     PROC SORT;
24        BY SS;
```

```
25      PROC PRINT;
26          TITLE 'STUDENT ROSTER';
27          ID SS;
28          VAR RAW PERCENT;
29      PROC MEANS MAXDEC=2 N MEAN STD RANGE MIN MAX;
30          TITLE 'CLASS STATISTICS';
31          VAR RAW PERCENT;
32      PROC CHART;
33          TITLE 'HISTOGRAM OF STUDENT SCORES';
34          VBAR PERCENT / MIDPOINTS=50 TO 100 BY 5;
35      PROC FREQ;
36          TITLE 'FREQUENCY DISTRIBUTION OF STUDENT ANS';
37          TABLES ANS1-ANS10;
```

Explanation of the program:

Line 3 instructs the program to "RETAIN" the variables KEY1-KEY10 for each observation. This is necessary since SAS programs will, by default, set all variables to **missing** before each observation is read. The new values then replace the missing values. Since the values for the KEY variables are read only once, the RETAIN statement prevents the program from setting the values to missing and instead uses the value from the previous observation. The answer key is read separately from the student answers by use of the "observation counter," which is designated by _N_. This is a "built-in" SAS variable that has the value of the present observation number. Thus, when _N_ is 1 we are reading the key. For all other values of _N_, we are reading student answers. Notice the DELETE statement in line 9. We include this since we do not want the first observation (the answer key) to be part of the output data set. The test scoring is done in lines 13 through 16. It simply states that if the student answer is equal to the answer key then the scoring variable should be set to 1, otherwise it is set to 0.

Since the raw score on a test is the number of correct responses, the SUM function in line 17 computes each student's raw score. The next line converts the raw score to a percentage score. Next we make use of the built-in format SSN11. to print the student social security numbers in standard format (this also ensures that the leading zeros in the number are printed).

The first several PROCs are straightforward. We want a student roster in SS number order (lines 23-28), the class statistics (lines 29-31), a histogram (lines 32-34), and the frequencies of 1's, 2's, etc. for each of the questions on the test (lines 35-37).

Selected output from these procedures is shown below:

```
STUDENT ROSTER

             SS     RAW     PERCENT

101-20-2333      7        70
102-01-0222      6        60
102-23-1324      7        70
112-34-6765      5        50
223-44-3232      9        90
453-54-5353      8        80

CLASS STATISTICS

N Obs  Variable  N    Minimum    Maximum      Range       Mean
-----------------------------------------------------------------
    6  RAW       6       5.00       9.00       4.00       7.00
       PERCENT   6      50.00      90.00      40.00      70.00
-----------------------------------------------------------------


N Obs  Variable      Std Dev
------------------------------------
    6  RAW             1.41
       PERCENT        14.14
------------------------------------
```

```
HISTOGRAM OF STUDENT SCORES

FREQUENCY BAR CHART
FREQUENCY

2 +                           ****
  |                           ****
  |                           ****
  |                           ****
  |                           ****
  |                           ****
  |                           ****
  |                           ****
  |                           ****
  |                           ****
1 +  ****         ****        ****        ****        ****
  |  ****         ****        ****        ****        ****
  |  ****         ****        ****        ****        ****
  |  ****         ****        ****        ****        ****
  |  ****         ****        ****        ****        ****
  |  ****         ****        ****        ****        ****
  |  ****         ****        ****        ****        ****
  |  ****         ****        ****        ****        ****
  |  ****         ****        ****        ****        ****
  |  ****         ****        ****        ****        ****
  ----------------------------------------------------------------
                                                                1
      5     5     6     6     7     7     8     8     9     9     0
      0     5     0     5     0     5     0     5     0     5     0
                              PERCENT MIDPOINT
```

FREQUENCY DISTRIBUTION OF STUDENT ANS

| ANS1 | Frequency | Percent | Cumulative Frequency | Cumulative Percent |
|------|-----------|---------|----------------------|--------------------|
| 1    | 4         | 66.7    | 4                    | 66.7               |
| 2    | 1         | 16.7    | 5                    | 83.3               |
| 3    | 1         | 16.7    | 6                    | 100.0              |

| ANS2 | Frequency | Percent | Cumulative Frequency | Cumulative Percent |
|------|-----------|---------|----------------------|--------------------|
| 1    | 1         | 16.7    | 1                    | 16.7               |
| 2    | 4         | 66.7    | 5                    | 83.3               |
| 3    | 1         | 16.7    | 6                    | 100.0              |

We can produce a compact table showing answer choice frequencies using PROC TABULATE. To do this efficiently, we will restructure the data set so that we will have a variable called QUESTION, which is the question number, and CHOICE, which is the answer choice for that question for each student. We will be fancy and create CHOICE as a character variable that shows the letter choice (A,B,C,D, or E) with an asterisk (*) next to the correct choice for each question. Again, we offer the program here without much explanation for those who might find the program useful or those who would like to figure out how it works. One of the authors (Smith) insists that good item analysis includes the mean test score for all students choosing each of the multiple choice items. Therefore, the code to produce this statististic is included as well. The details of TABULATE are too much to describe here and we refer you to either the **SAS User's Guide: Basics** or the **SAS Procedures Guide for Personal Computers** listed in Chapter 1 of this book.

The complete program to restructure the data set and produce the statistics described above is shown next:

```
1.     *PROGRAM TO SCORE A TEST;
2.     OPTIONS LS=64 PS=59 NOCENTER;
3.     PROC FORMAT;
4.        PICTURE PCT LOW-<0=' ' 0-HIGH='00000%';
5.     DATA SCORE;
6.     LENGTH Q1-Q10 $ 2;
7.     RETAIN KEY1-KEY10;
8.     ARRAY KEY{*} KEY1-KEY10;
9.     ARRAY ANS{*} ANS1-ANS10; *ANS ARE STUDENT ANS;
10.    ARRAY Q{*} $ Q1-Q10; *CHARACTER VARS USED WITH PROC TABULATE;
11.    ARRAY S{*} S1-S10;   *THE S'S ARE SCORED RESPONSES;
```

```
12.   IF _N_ = 1 THEN DO;
13.       INPUT @11 (KEY1-KEY10)(1.);
14.       DELETE;
15.       END;
16.   ELSE DO;
17.       INPUT SS 1-10 (ANS1-ANS10)(1.);
18.       DO I = 1 TO 10;
19.           Q{I} = LEFT(TRANSLATE (ANS{I},'ABCDE','12345'));
20.           IF KEY{I} = ANS{I} THEN DO;
21.               S{I}=1;
22.               DUMMY=Q{I};
23.               SUBSTR(DUMMY,2,1)='*';
24.               Q{I}=DUMMY;
25.               END;
26.           ELSE S{I}=0;
27.           END;
28.       RAW=SUM (OF S1-S10);
29.       PERCENT=100*RAW/10;
30.       DROP I KEY1-KEY10;
31.       END;
32.   FORMAT SS SSN11.;
33.   CARDS;
ANS KEY   1234554321
102231324 1234323321
101202333 3334254321
102010222 1233544211
123456789 1234112345
343212323 2213554115
      etc.
34.   DATA TEMP;
35.   SET SCORE;
36.   KEEP QUESTION CHOICE PERCENT;
37.   ARRAY Q{*} $ 2 Q1-Q10;
38.   DO QUESTION=1 TO 10;
39.       CHOICE=Q{QUESTION};
40.       OUTPUT;
41.       END;
42.   PROC TABULATE;
43.       CLASS QUESTION CHOICE;
44.       VAR PERCENT;
45.       TABLE QUESTION*CHOICE
46.       ,PERCENT=' '*(PCTN<CHOICE>*F=PCT. MEAN*F=PCT.
47.       STD*F=10.2)   / RTS=20 MISSTEXT=' ';
48.       KEYLABEL ALL='TOTAL' MEAN='MEAN SCORE' PCTN='FREQ'
49.                 STD='STANDARD DEVIATION';
```

A brief explanation of the program follows:

Lines 3 and 4 create a format to be used with PROC TABULATE so that percentage scores will be printed with the "%" sign. Lines 5 through 18 are the same as the previous program with the exception of the Q variables. The Q's are character variables of length 2 which hold the student's answer choice. The first byte of this variable is a letter A,B,C,D, or E. The second byte is either a blank or an asterisk (*), representing the correct answer for a particular question. Line 19 uses the TRANSLATE function which translates the numerical responses (1,2,3,4,5) to letters (A,B,C,D,E) and left justifies it (with the LEFT function). Line 23 places an asterisk in the second position of the correct answer choice. (Note: Lines 22 and 24 are necessary for version 6.02 of SAS/PC because of a minor bug. It is expected that all versions later than 6.02 will have this problem corrected, in which case, the variable Q{I} will be placed directly in the SUBSTR pseudofunction.) Lines 34 through 41 restructure the data set as mentioned earlier. This data set contains n observations per student, where n is the number of items on the test. Careful, you cannot use this data set to compute the test standard deviation! The original data set must be used for this purpose.

Selected portions of the output from these procedures are shown in the tables below:

| QUESTION | CHOICE | FREQ | MEAN SCORE | STANDARD DEVIATION |
|----------|--------|------|-----------|--------------------|
| 1 | A* | 36% | 52% | 23.60 |
| | B | 15% | 30% | 10.00 |
| | C | 15% | 53% | 37.86 |
| | D | 15% | 16% | 5.77 |
| | E | 15% | 33% | 23.09 |
| 2 | CHOICE | | | |
| | A | 5% | 20% | |
| | B* | 47% | 54% | 22.97 |
| | C | 10% | 45% | 35.36 |
| | D | 15% | 33% | 23.09 |
| | E | 21% | 17% | 9.57 |
| 3 | CHOICE | | | |
| | A | 10% | 30% | 14.14 |
| | B | 21% | 20% | 8.16 |
| | C* | 36% | 64% | 13.97 |
| | D | 21% | 35% | 31.09 |
| | E | 10% | 20% | 0.00 |

etc.

The frequency column shows the percentage of students selecting each item choice. The frequency next to the correct answer (marked by an *) is the item difficulty (percent of students answering the item correctly). The column labeled MEAN SCORE shows the mean test score for all students answering the answer choice listed to the left. For example, for item 1, 23.6% of the students chose A, which is the correct answer. The students who chose A had a mean score of 36% on the test. The mean score of all students who chose B was 15%, and so forth.

Computing test reliability is shown next. This program computes a test statistic called coefficient alpha, which, for a test item that is dichotomous, is equivalent to the Kuder-Richardson formula 20. The key here is to output a data set that contains the item and test **variances**. Here is the program:

```
1        PROC MEANS NOPRINT DATA=SCORE;
2            VAR S1-S10 RAW;
3            OUTPUT OUT=VAROUT VAR=VS1-VS10 VRAW;
4        DATA _NULL_ ;
5        FILE PRINT;
6        SET VAROUT;
7        SUMVAR = SUM (OF VS1-VS10);
8        KR20 = (10/9)*(1-SUMVAR/VRAW);
9        PUT KR20= ;
```

As we discussed in Reference Chapter 2, PROC MEANS can output data sets containing various statistics. Here, the keyword VAR= is used to output a data set of item variances and test variance. This data set contains only one observation. In order to sum the item variances, we need to use another data step. You may not be familiar with the special SAS data set name _NULL_ in line 4. This reserved data set name instructs the SAS system to process the observations as they are encountered but **not** to write them to a temporary or permanent SAS data set. This saves time and--if you use a mainframe computer--money. Line 7 computes the sum of the item variances and line 8 is the formula for coefficient alpha. The formula is

$$ALPHA = \frac{k}{k-1} \times \frac{(1-SUM(\text{item variances}))}{\text{test variance}}$$

where k is the number of items on the test.

We get the program to print the results for us by using a PUT statement (line 9). The results of this PUT are sent to the OUTPUT (window if you are using SAS/PC) because of the FILE PRINT statement in line 5.

## B. READING "UNSTRUCTURED" DATA (STREAM INPUT)

Almost all the examples we have shown you so far have been either small data sets or balanced data sets that were relatively easy to read using standard INPUT statements. However, in the "real" world, we often encounter data sets that are not so "clean." For example, we might have a varying number of records for each subject in a study. Another example would be an unbalanced design where there were different numbers of subjects in each treatment. As these data sets become large, reading them without error sometimes becomes the most difficult part of the data processing problem. The techniques shown in this section will allow you to read almost any type of unstructured data easily.

The key to all the examples that follow is to imbed "tags" in the data to indicate to the program what type of data to read. A t-test example with unequal n's and an unbalanced ANOVA will serve to illustrate the use of tags and stream data input.

### Example 1--Unbalanced T-test.

The amount and complexity of the data have been reduced to make the examples short and easy to follow. The strength of the techniques is their use with larger, more complicated data sets.

We want to analyze an experiment where we had 5 control and 3 treatment subjects and we recorded a single variable per subject. The data are shown below:

```
                          GROUP
    ------------------------------------------
    CONTROL                          TREATMENT
        20                               40
        25                               42
        23                               35
        27
        30
```

The simplest, most straightforward method to read these data is shown next:

**Example 1-A.**

```
*TRADITIONAL INPUT METHOD;
DATA EX1A;
INPUT GROUP $ X @@;
CARDS;
C 20 C 25 C 23 C 27 C 30
T 40 T 42 T 35
PROC TTEST;
   CLASS GROUP;
   VAR X;
```

For larger amounts of data, this program has some problems. It is tedious and time consuming to repeat the group identification before each variable to be read. This can be corrected in two ways: First, we can put the information concerning the number of observations per group in the **program** (Example 1-B) or we can put this information in the **data** itself (Example 1-C). As mentioned above, if the number of observations were large (several hundred or more), a single mistake in counting would have disastrous consequences.

**Example 1-B.**

```
DATA EX1B;
GROUP='C';
DO I=1 TO 5;
    INPUT X @@;
    OUTPUT;
    END;
GROUP='T';
DO I=1 TO 3;
    INPUT X @@;
    OUTPUT;
    END;
DROP I;
CARDS;
20 25 23 27 30
40 42 35
PROC TTEST;
    CLASS GROUP;
    VAR X;
```

**EXAMPLE 1-C.**

```
DATA EX1C;
DO GROUP='C','T';
    INPUT N;
    DO I=1 TO N;
        INPUT X @@;
        OUTPUT;
        END;
    END;
DROP N I;
CARDS;
5
20 25 23 27 30
3
40 42 35
PROC TTEST;
    CLASS GROUP;
    VAR X;
```

The method we are suggesting for large data sets is shown in Example 1-D below:

**Example 1-D.**

```
*READING THE DATA WITH TAGS;
DATA EX1D;
RETAIN GROUP;
INPUT TEST $ @@;
DO WHILE (TEST='C' OR TEST='T');
    GROUP=TEST;
    RETURN;
    END;
X=INPUT (TEST,5.0);
OUTPUT;
DROP TEST;
CARDS;
C 20 25 23 27 30
T 40 42 35
PROC TTEST;
    CLASS GROUP;
    VAR X;
```

With this program we can add or delete data without making any changes to our program. The two important points in the program are

1.   All  data  items  are read  as  **character**  and interpreted.

2. The **INPUT function** is used to "reread" the data with a numeric format.

This same program can read data that are not as ordered as Example 1-D. For instance, the data set

```
C 20 25 23 T 40 42
C 30 T 35
```

will also be read correctly.  For large data sets,  this structure  is  less  prone to error  than  Examples  1-A through 1-C.  (Of course,  we pay additional  processing costs  for the alternative program but the ease of  data entry and the elimination of counting errors is probably worth the extra cost.)

### Example 2--Unbalanced Two-Way ANOVA.

The  next example will be an unbalanced design  for which  we want to perform an analysis of variance.  Our design is as follows:

```
                          GROUP

             A                B               C
      -------------------------------------------------
            20               70              90
            30               80              90
      M     40               90              80
            20                               90
            50
SEX   -------------------------------------------------
            25               70              20
            30               90              20
      F     45               90              30
            30               80
            65               85
            72
```

The  straightforward method of entering these  data would be

```
DATA EX2A;
INPUT GROUP $ SEX $ SCORE;
CARDS;
A M 20
A M 30
    etc.
```

This  is a lengthy and wasteful data entry  method.
For  small data sets of this type,  we could follow  the
example  of the unbalanced t-test problem and enter  the
number  of observations per cell,  either in the program
or imbedded in the data. A preferable method, especially
for a large number of observations per cell where counting  would  be inconvenient,  is shown  in  Example  2-B
below:

```
*FIRST METHOD OF READING ANOVA DATA WITH TAGS;
DATA EX2A;
DO SEX='M','F';
    DO GROUP='A','B','C';
       INPUT TEST $ @;
       DO WHILE (TEST NE '#');
          SCORE=INPUT(TEST,6.0);
          OUTPUT;
          INPUT TEST $ @;
          END;
       END;
    END;
DROP TEST;
CARDS;
20 30 40 20 50 # 70 80 90
# 90 90 80 90 # 25 30 45 30
65 72 # 70 90 90 80 85 # 20 20 30 #
PROC GLM;
    etc.
```

This  program reads and assigns observations  to  a
cell until a "#" is read in the data stream. The program
then  finishes the innermost loop and the next  cell  is
selected. We can read as many lines as necessary for the
observations for a given cell.

An  improved version of this program is shown  next
(Example 2-B). With this program,  we can read the cells
in  any order and do not have to supply the program with
the  cell identification since it is incorporated  right

in the tags. Let's look over the program first, and then
we will discuss the salient features:

```
*MORE ELEGANT METHOD FOR UNBALANCED ANOVA DESIGN;
DATA EX2B;
RETAIN GROUP SEX;
INPUT TEST $ @@;
IF VERIFY (TEST,'ABCMF ') = 0 THEN DO;
   GROUP = SUBSTR (TEST,1,1);
   SEX = SUBSTR (TEST,2,1);
   DELETE;
   RETURN;
   END;
SCORE = INPUT (TEST,6.);
DROP TEST;
CARDS;
AM 20 30 40 20 50
BM 70 80 90
CM 90 80 80 90
AF 25 30 45 30 65 72
BF 70 90 90 80 85
CF 20 20 30
PROC GLM;
   etc.
```

This program allows us to enter the cells in any
order and even use as many lines as necessary for the
observations from a cell. This form of data entry is
also convenient when we will be adding more data at a
later time. The analysis can be rerun without any
changes to the program. Additional observations can even
be added at the end of the original data.

Special features of this program are the use of the
**VERIFY** and **SUBSTR** functions. The **VERIFY** function returns
0 if all the characters of the the variable TEST can be
found as one of the characters in the second argument of
the function. Note that a blank is included in argument
2 of the VERIFY function since the length of TEST is, by
default, equal to 8 bytes, which means that it will contain two letters and 6 blanks. The SUBSTR function picks
off the GROUP and SEX values from the TEST string and
the INPUT function converts all character values back to
numeric.

# Problem Sets

Answers to **odd**-numbered problems are provided following the exercises. Problems marked with an asterisk (*) are especially challenging.

**CHAPTER 1 PROBLEMS:**

1-1 We have collected the following data on 5 subjects:

| ID | AGE | SEX | GRADE POINT AVERAGE (GPA) | COLLEGE ENTRANCE EXAM SCORE (CSCORE) |
|----|-----|-----|---------------------------|--------------------------------------|
| 1  | 18  | M   | 3.7 | 650 |
| 2  | 18  | F   | 2.0 | 490 |
| 3  | 19  | F   | 3.3 | 580 |
| 4  | 23  | M   | 2.8 | 530 |
| 5  | 21  | M   | 3.5 | 640 |

(a) Write the necessary SAS statements to create a SAS data set.

(b) Add the statement(s) necessary to compute the mean grade point average and mean college entrance exam score.

(c) We want to compute an index for each subject, as follows:

$$INDEX = GPA + 3 \times CSCORE/500$$

Modify your program to compute this INDEX for each student and to print a list of students in order of increasing INDEX. Include in your listing the student ID, GPA, CSCORE, and INDEX.

1-2 Given the following set of data:

| Social security number | Annual salary | Age | Race |
|------------------------|---------------|-----|------|
| 123874414 | 28,000 | 35 | W |
| 646239182 | 29,500 | 37 | B |
| 012437652 | 35,100 | 40 | W |
| 018451357 | 26,500 | 31 | W |

(a)   Write  a SAS program that  will  compute  the average annual salary and age.

(b)   If  all subjects were in a 30% tax  bracket, compute  their  taxes (based on gross salary)   and print out a list,  in social security number order, showing the annual salary and the tax.

1-3 What's wrong with this program?

```
DATA MISTAKE;
    INPUT ID 1-3 TOWN 4-6 REGION 7-9 YEAR 11-12 BUDGET 12-14
    VOTER TURNOUT 16-20
(data cards go here)
PROC MEANS;
    VAR ID REGION VOTER TURNOUT;
    N,STD,MEAN;
```

*1-4 A  large corporation is interested in who is buying their  product.  What the CEO wants is a profile of  the "typical buyer."  The variables collected on a sample of buyers are:  age, sex, race, income, marital status, and homeowner/renter.   Set up a layout for the observations and  write  a SAS program to get  the  profile  of  the "typical buyer."

Hints and Comments:

(1)  For variables such as "homeowner,"  it is easier to remember  what you have if you let negative responses be 0 and positive responses be 1.

(2)   When grouping numerical variables into categories, make  sure  your grouping fits your needs and your data. For  example,  if your product was denture  cream,   the grouping of age (1=<21  2=21-35 3=36-50 4=>50)  would be nearly  useless.  You know these people are mostly  over 50.   You might want groupings such as (1=<50,   2=50-59 3=60-69 4=>69).

1-5 Given the data set

| ID | RACE | SBP | DBP | HR |
|-----|------|-----|-----|-----|
| 001 | W | 130 | 80 | 60 |
| 002 | B | 140 | 90 | 70 |
| 003 | W | 120 | 70 | 64 |
| 004 | W | 150 | 90 | 76 |
| 005 | B | 124 | 86 | 72 |

Write the SAS statements to produce a report as follows:

RACE AND HEMODYNAMIC VARIABLES

| ID | RACE | SBP | DBP |
|-----|------|-----|-----|
| 003 | W | 120 | 70 |
| 005 | B | 124 | 86 |
| 001 | W | 130 | 80 |
| 002 | B | 140 | 90 |
| 004 | W | 150 | 90 |

Note: 1. There is no "OBS" column.
2. Data is in increasing order of SBP.
3. The variable HR is not included in the report.
4. The report has a title.

**CHAPTER 2 PROBLEMS:**

2-1 Add the necessary statements to compute the number of males and females in Problem 1-1.

2-2 Given the data set from Problem 1-2, use SAS to compute the number of Whites(W) and Blacks(B).

2-3 We have a SAS data set containing variables X, Y, Z, and GROUP.

(a) Write the SAS statements to generate a frequency bar chart (histogram) for GROUP (assume GROUP is a categorical variable).

(b)  Write the SAS statements to generate a plot of
Y vs.  X (with "Y" on the vertical axis and "X"  on
the horizontal).

(c) Write the SAS statements to generate a separate
plot  of  Y  vs.   X  for each value of  the  GROUP
variables.

2-4 We have recorded the following data from an
experiment:

```
SUBJECT   DOSE REACT LIVER_WT SPLEEN
-----------------------------------
   1       1    5.4    10.2    8.9
   2       1    5.9     9.8    7.3
   3       1    4.8    12.2    9.1
   4       1    6.9    11.8    8.8
   5       1   15.8    10.9    9.0
   6       2    4.9    13.8    6.6
   7       2    5.0    12.0    7.9
   8       2    6.7    10.5    8.0
   9       2   18.2    11.9    6.9
  10       2    5.5     9.9    9.1
```

Use  PROC  UNIVARIATE  to  produce  histograms,   normal
probability plots, and box plots, and test the distributions for normality.  Do this  for the variables  REACT,
LIVER_WT,  and SPLEEN,  first for all subjects and  then
separately for each of the two DOSES.

2-5 What's wrong with this program?

```
DATA;
INPUT AGE STATUS PROGNOSIS DOCTOR SEX STATUS2
      STATUS3;
(data cards)
PROC CHART BY SEX;
   VBAR STATUS
   VBAR PROGNOSIS;
PROC PLOT;
   DOCTOR BY PROGNOSIS;
```

2-6 Given the data set

| Salesperson | Target company | Number of visits | Number of phone calls | Units sold |
|---|---|---|---|---|
| Brown | American | 3 | 12 | 28,000 |
| Johnson | VRW | 6 | 14 | 33,000 |
| Rivera | Texam | 2 | 6 | 8,000 |
| Brown | Standard | 0 | 22 | 0 |
| Brown | Knowles | 2 | 19 | 12,000 |
| Rivera | Metro | 4 | 8 | 13,000 |
| Rivera | Uniman | 8 | 7 | 27,000 |
| Johnson | Oldham | 3 | 16 | 8,000 |
| Johnson | Rondo | 2 | 14 | 2,000 |

(a)  Write a SAS program to compare the sales records of the company's three salespeople.

(b)  Plot the number of visits against the number of phone calls. Use "Salesperson" as the plotting symbol (instead of the usual A, B, C etc.).

(c)  Make a frequency bar chart for each salesperson for the variable "units sold."

*2-7  You have completed an experiment and recorded a subject ID, and values for variables A, B, and C. You want to compute means for A, B, and C but, unfortunately, your lab technician, who didn't know SAS programming, arranged the data like this:

| ID | TYPE | SCORE |
|---|---|---|
| 1 | A | 44 |
| 1 | B | 9 |
| 1 | C | 203 |
| 2 | A | 50 |
| 2 | B | 7 |
| 2 | C | 188 |
| 3 | A | 39 |
| 3 | B | 9 |
| 3 | C | 234 |
| | etc. | |

Write a program to read this data set and produce means. (Hint: Remember the power of BY variable processing.)

## CHAPTER 3 PROBLEMS:

3-1  Suppose we have a variable called GROUP that has numeric values of 1,2, or 3. Group 1 is a control group, group 2 is given aspirin, and group 3 is given Tylenol. Create a format to be assigned to the GROUP variable.

3-2  A survey was conducted and data were collected and coded. The data layout is shown below (all values are numeric):

```
VARIABLE  DESCRIPTION          COLUMNS  CODING VALUES
-------------------------------------------------------
ID        Subject identifier    1-3

SEX                              4       1=Male 2=Female

PARTY     Political party        5       1=Republican
                                         2=Democrat
                                         3=Not registered

VOTE      Did you vote in the    6       0=No 1=Yes
          last election?

FOREIGN   Do you agree with the  7       0=No 1=Yes
          government's foreign
          policy?

SPEND     Should we increase     8       0=No 1=Yes
          domestic spending?
```

Collected data are shown below:

```
00711110
01322101
13721001
117 1111
42813110
01723101
03712101
```

(a) Create a SAS data set, complete with **labels** and **formats** for this questionnaire.

(b)   Generate frequency counts for the variables SEX, PARTY, VOTE, FOREIGN, and SPEND.

(c)   Test if there is a relationship between voting in the last election versus agreement with spending and foreign policy.   (Have SAS compute  chi-square for these relationships.)

3-3  We  have a SAS data set  containing  the  variables WEIGHT,  HEIGHT, SEX, and RACE. We want to recode WEIGHT and HEIGHT as follows:

```
WEIGHT     0-100   = 1
           101-150 = 2
           151-200 = 3
             >200  = 4

HEIGHT     0-70    = 1
             >70   = 2
```

We then want  to  generate  a  table of WEIGHT cat-egories  (rows)  by HEIGHT categories (columns).  Recode these variables in two  ways:   First,  with "IF" state-ments;  second with format  statements.   Then write the necessary  statements to generate the table.

*3-4  A  physical exam was given to a group of patients. Each  patient was diagnosed to have none, 1,  2,  or 3 problems from the code list below:

```
     CODE        PROBLEM DESCRIPTION
-------------------------------------------
      1          Cold
      2          Flu
      3          Trouble sleeping
      4          Chest pain
      5          Muscle pain
      6          Headaches
      7          Overweight
      8          High blood pressure
      9          Hearing loss
```

The coding scheme is as follows:

```
VARIABLE   DESCRIPTION                          COLUMN(S)
-------------------------------------------------------
SUBJ       Subject number                       1-2
PROB1      Problem 1                            3
PROB2        "    2                             4
PROB3        "    3                             5
HR         Heart rate                           6-8
SBP        Systolic blood pressure              9-11
DBP        Diastolic blood pressure             12-14
```

Using the sample data below:

```
   COLUMNS      12345678901234
------------------------------------
               11127 78130 80
               1787  82180110
               031   62120 78
               4261  68130 80
               89    58120 76
               9948  82178100
```

(a) Compute the mean HR, SBP, and DBP.

(b) Generate frequency counts for each of the nine medical problems.

3-5 What's wrong with this program?

```
1       DATA IGOOFED;
2       INPUT #1 ID 1-3 SEX 4 AGE 5-6 RACE 7(QUES1-QUES10)(1.)
            #2 @4 (QUES11-QUES25) (1.);
3       FORMAT SEX SEX. RACE RACE. QUES1-QUES25 YESNO.;
4       CARDS;
        00112311010011101
          11001110011010
        00224421011101110
          01110111011110
            etc.
5       PROC FORMAT;
6          VALUE SEX 1='MALE' 2='FEMALE';
7          VALUE RACE 1='WHITE' 2='BLACK' 3='HISPANIC';
8          VALUE YESNO 0='NO' 1='YES';
9       PROC FREQ;
10         VAR SEX RACE QUES1-QUES25 / CHISQ;
11      PROC MEANS MAXDEC=2 N MEAN STD MIN MAX;
12         BY RACE;
13         VAR AGE;
```

Hints and comments:  The INPUT statement is correct. The
pointers (@ signs) and format lists (1.)  are described
in Reference Chapter 2, Section F. There are 4 errors.


**CHAPTER 4 PROBLEMS:**

4-1 Given the following data:

```
      X      Y    Z
--------------------
      1      3   15
      7     13    7
      8     12    5
      3      4   14
      4      7   10
```

(a)    Write  a SAS program and compute the  Pearson
correlation coefficient between X and Y;  X  and Z.
What is the significance of each? (Note, for SAS/PC
version  6.02,   only a correlation matrix  can  be
produced.)

(b)    Change  the correlation request to produce  a
correlation matrix, i.e.,   the correlation coefficient  between  each variable against  every  other
variable.

4-2 Given the following data:

```
        AGE          SYSTOLIC BLOOD PRESSURE
-------------------------------------------
        15                 116
        20                 120
        25                 130
        30                 132
        40                 150
        50                 148
```

How  much  of  the  variance  of  SBP  (systolic  blood
pressure)  can  be explained by the fact that  there  is
variability in AGE?   (Use SAS to compute the correlation
between SBP and AGE.)

4-3 From the data for X and Y in Problem 4-1:

(a) Use SAS to compute a regression line (Y on X).

(b) What is the slope and intercept?

(c) Are they significantly different from zero?

4-4 Using the data from 4-1, compute three new variables LX, LY, and LZ, which are the natural logs of the original values. Compute a correlation matrix for the three new variables.

4-5 Generate

(a) a plot of Y vs. X (data from 4-1);

(b) a plot of the regression line and the original data on the same set of axes.

4-6 Given the data set

| COUNTY | POP | HOSPITAL | FIRE_CO | RURAL |
|--------|-----|----------|---------|-------|
| 1 | 35 | 1 | 2 | YES |
| 2 | 88 | 5 | 8 | NO |
| 3 | 5 | 0 | 1 | YES |
| 4 | 55 | 3 | 3 | YES |
| 5 | 75 | 4 | 5 | NO |
| 6 | 125 | 5 | 8 | NO |
| 7 | 225 | 7 | 9 | YES |
| 8 | 500 | 10 | 11 | NO |

(a) Write a SAS program to create a SAS data set of the above data.

(b) Run PROC UNIVARIATE to check the distributions for the variables POP, HOSPITAL, and FIRE_CO.

(c) Compute a correlation matrix for the variables POP, HOSPITAL, and FIRE_CO. Produce both Pearson and Spearman correlations. Which is more appropriate?

(d) Recode POP, HOSPITAL, and FIRE_CO so that they each have two levels (use a median cut or a value somewhere near the 50th percentile). Compute cross-tabulations between the variable RURAL and the recoded variables.

4-7 What's wrong with this program?

```
1       DATA MANY-ERR;
2       INPUT X Y Z;
3       IF X LE 0 THEN X=1;
4       IF Y LE 0 THEN Y=1;
5       IF Z LE 0 THEN Z=1;
6       LOGX=LOG(X);
7       LOGY=LOG(Y);
8       LOGZ=LOG(Z);
9       CARDS;
        1 2 3
        . 7 8
        4 . 10
        7 8 11
10      PROC CORR / PEARSON SPEARMAN;
11          VAR X-LOGZ;
```

**CHAPTER 5 PROBLEMS:**

5-1 The following table shows the time (in minutes) for subjects to feel relief from headache pain:

(Cure time in minutes)

| Aspirin | Tylenol |
|---------|---------|
| 40 | 35 |
| 42 | 37 |
| 48 | 42 |
| 35 | 22 |
| 62 | 38 |
| 35 | 29 |

Write a SAS program to read these data and perform a t-test. Is either product significantly quicker than the other (at the .05 level)?

5-2 Using the same data as 5-1, perform a Wilcoxon rank-sum test and a median test.

5-3 In another study, 4 subjects are given Drug A for a headache and then Drug B the next time they have a headache. Four other subjects are given Drug B first, and Drug A for their second headache. Again, the "cure" time is recorded. What type of test will you use to test if one drug is faster than the other? Below are some made-up data: Write the SAS statements to run the appropriate analysis.

```
Subject       Drug A   Drug B
------------------------------
   1            20        18
   2            40        36
   3            30        30
   4            45        46
   5            19        15
   6            27        22
   7            32        29
   8            26        25
```

*5-4 A researcher wants to randomly assign 30 patients to one of three treatment groups. Each subject has a unique subject number (SUBJ). Write a SAS program to assign these subjects to a treatment group.

5-5 What's wrong with this program?

```
1       DATA DRUGSTDY;
2       INPUT SUBJ 1-3 DRUG 4 HEARTRATE 5-7 SBP 8-10
3           DBP 11-13;
4       AVEBP=DBP + (SBP-DBP)/3;
5       CARDS;
        0011064130080
        0021068120076
        0031070156090
        0042080140080
        0052088180092
        0062098178094
```

```
6          PROC NPAR1WAY WILCOXON MEDIAN;
7              TITLE 'MY DRUG STUDY';
8              CLASS DRUG;
9              VAR HEARTRATE SBP DBP AVEBP;
10         PROC T-TEST;
11             CLASS DRUG;
12             VAR HEARTRATE SBP DBP AVEBP;
```

**CHAPTER 6 PROBLEMS:**

6-1 The next two questions were inspired by one of the authors (Cody) watching the French Open Tennis tournament while working on problem sets. (McEnroe versus Lendl (1984). Lendl won in 5 sets.)

Three brands of tennis shoes were tested to see how many months of playing would wear out the sole. Eight pairs of brands A, N, and T were randomly assigned to a group of 24 volunteers. The table below shows the results of the study:

|  | BRAND | |  |
|---|---|---|---|
|  | A | N | T |
| | 8 | 4 | 12 |
| | 10 | 7 | 8 |
| Wear time | 9 | 5 | 10 |
| in months | 11 | 5 | 10 |
| | 10 | 6 | 11 |
| | 10 | 7 | 9 |
| | 8 | 6 | 9 |
| | 12 | 4 | 12 |

Are the brands equal in wear quality? Write a SAS program to solve this problem, using ANOVA.

6-2 Tennis balls are tested in a machine to see how many bounces they can withstand before they fail to bounce 30% of their dropping height. Two brands of balls (W and P) are compared. In addition, the effect of shelf life on these brands is tested. Half of the balls of each brand are 6 months old, the other half, fresh. Using a two-way analysis of variance, what conclusions can you reach? The data are shown below:

```
                              BRAND
                  W          |         P
            ---------------- | ----------------
                 6.7         |        7.5
                 7.2         |        7.6
        NEW      7.4         |        8.0
                 8.2         |        7.2
                 8.1         |        7.3
AGE  --------------------------------------------
                 4.6         |        6.3
                 4.4         |        6.2
        OLD      4.5         |        6.6
                 5.1         |        6.2
                 4.3         |        6.0
```

6-3  A  taste test was conducted to rate the  preference between  brands  C and P of a  popular  soft  drink.  In addition,  the  age category (1= less than 20,  2= 20 or more) was recorded. Preference data (on a scale of 1-10) are displayed below:

```
                              BRAND
                  C          |         P
            ---------------- | ----------------
                  7          |         9
                  6          |         8
                  6          |         9
       <20        5          |         9
                  6          |         8
AGE  --------------------------------------------
                  9          |         6
                  8          |         7
                  8          |         6
       >=20       9          |         6
                  7          |         5
                  8          |
                  8          |
```

(a)  Write a SAS program to analyze these data with a  two-way analysis of variance.  (Careful:  is the design balanced?)

(b) Draw an interaction graph.

(c) Follow  up with a t-test comparing brand  C  to brand P for each age group separately.

6-4   A  manufacturer wants to reanalyze  the  data  in
Problem  6-1,  omitting all data for brand N.   Run  the
appropriate analysis.


6-5 What's wrong with this program?

```
1       DATA TREE;
2       INPUT TYPE $ LOCATION $ HEIGHT;
3       CARDS;
        PINE NORTH 35
        PINE NORTH 37
        PINE NORTH 41
        PINE NORTH 41
        MAPLE NORTH 44
        MAPLE NORTH 41
        PINE SOUTH 53
        PINE SOUTH 55
        MAPLE SOUTH 28
        MAPLE SOUTH 33
        MAPLE SOUTH 32
        MAPLE SOUTH 22
4       PROC ANOVA;
5          CLASSES TYPE LOCATION;
6          MODEL HEIGHT = TYPE|LOCATION;
7          MEANS TYPE LOCATION TYPE*LOCATION;
```

**CHAPTER 7 PROBLEMS:**

A marketing survey is conducted to determine sport shirt
preference.  A  questionnaire is administered to a panel
of  4 judges.  Each judge rates three shirts of each  of
three  brands (X,  Y,  and Z).   The data entry form  is
shown below:

### MARKETING SURVEY FORM

```
1. Judge ID                        |__|1
2. Brand (1=X, 2=Y, 3=Z)           |__|2
3. Color rating 9=Best, 1=Worst    |__|3
4. Workmanship rating              |__|4
5. Overall preference              |__|5
```

An index is computed as follows:

INDEX = (3*OVERALL PREFERENCE + 2*WORKMANSHIP +
         COLOR RATING) / 6.0

The collected data follow:

```
11836
21747
31767
41846
12635
22534
32546
42436
13988
23877
33978
43887
11758
21755
31847
41756
12464
22545
32455
42554
13786
23889
33976
43879
```

Compare the color rating, workmanship, overall preference, and index among the three brands, using analysis of variance. (Hint: This is a repeated measures design--each judge rates all three brands.)

7-2 A taste test is conducted to see which city has the best-tasting tap water. A panel of 4 judges tastes each of the samples from the four cities represented. The rating scale is a Likert scale with 1=worst to 9=best. Sample data and the coding scheme are shown below:

```
     COLUMN              DESCRIPTION
----------------------------------------------------------
     1-3                 Judge identification number
     4                   City code:
                         1=New York 2=New Orleans
                         3=Chicago 4=Denver
     5                   Taste rating 1=worst 9=best
```

Data:

```
00118
00126
00138
00145
00215
00226
00235
00244
00317
00324
00336
00344
00417
00425
00437
00443
```

Write a SAS program to describe these data and to perform an analysis of variance. Remember that we have a repeated measures design.

*7-3  The same data as in Problem 7-2 are to be analyzed. However, they are arranged so that the four ratings from each judge are on one line. Thus, columns 1-3 are for the judge ID, column 4 is the rating for New York, column 5 for New Orleans, column 6 for Chicago, and column 7 for Denver. Our reformed data are shown below:

```
0018685
0025654
0037464
0047573
```

Write the DATA statements to analyze this arrangement of the data. Remember, you will need to create a variable for CITY and to have one observation per city.

*7-4  A study is conducted to test the area of nerve fibers in NORMAL and DIABETIC rats. A sample from the DISTAL and PROXIMAL ends of each nerve fiber is measured for each rat. Therefore, we have GROUP (Normal versus CONTROL) and LOCATION (Distal versus proximal) as independent variables, with location as a repeated measure (each rat nerve is measured at each end of the nerve fiber). The data are shown below:

|          | RATNO | DISTAL | PROXIMAL |
|----------|-------|--------|----------|
| NORMAL   | 1     | 34     | 38       |
|          | 2     | 28     | 38       |
|          | 3     | 38     | 48       |
|          | 4     | 32     | 38       |
| DIABETIC | 5     | 44     | 42       |
|          | 6     | 52     | 48       |
|          | 7     | 46     | 46       |
|          | 8     | 54     | 50       |

Write a SAS program to enter these data and run a two-way analysis of variance, treating the location as a repeated measure. Use the REPEATED option for the LOCATION variable. Is there any difficulty in interpreting the main effects? Why?

7-5 What's wrong with this program?

```
1      DATA FINDIT;
2      DO GROUP='CONTROL','DRUG';
3         DO TIME='BEFORE','AFTER';
4            DO SUBJ=1 TO 3;
5               INPUT SCORE @;
6               END;
7            END;
8         END;
9      CARDS;
      10 13 15 20      (data for subject 1)
      12 14 16 18      (data for subject 2)
      15 18 22 28      (data for subject 3)
10     PROC ANOVA;
11        TITLE 'ANALYSIS OF VARIANCE';
12        CLASSES SUBJ GROUP TIME;
13        MODEL SCORE = GROUP SUBJ(GROUP)
14           TIME GROUP*TIME TIME*SUBJ(GROUP);
15        TEST H=GROUP E=SUBJ(GROUP);
16        TEST H=TIME GROUP*TIME E=TIME*SUBJ(GROUP);
17        MEANS GROUP|TIME;
```

Note: the comments in parentheses are not part of the program.

**CHAPTER 8 PROBLEMS:**

8-1 We want to test the effect of light level and amount of water on the yield of tomato plants. Potted plants receive 3 levels of light and 2 levels of water. The yield, in pounds, is recorded. The results are as follows:

| YIELD | LIGHT | WATER | | YIELD | LIGHT | WATER |
|---|---|---|---|---|---|---|
| 12 | 1 | 1 | | 20 | 2 | 2 |
| 9 | 1 | 1 | | 16 | 2 | 2 |
| 8 | 1 | 1 | | 16 | 2 | 2 |
| 13 | 1 | 2 | | 18 | 3 | 1 |
| 15 | 1 | 2 | | 25 | 3 | 1 |
| 14 | 1 | 2 | | 20 | 3 | 1 |
| 16 | 2 | 1 | | 25 | 3 | 2 |
| 14 | 2 | 1 | | 27 | 3 | 2 |
| 12 | 2 | 1 | | 29 | 3 | 2 |

Write a SAS program to read these data and perform a multiple regression (use PROC REG).

8-2 We want to estimate the number of books in a college library. Data are collected from colleges across the country of the number of volumes, the student enrollment (in thousands), the highest degree offered (1=B.A., 2=M.A., 3=Ph. D.), and size of the main campus (in acres). Results of this (hypothetical) study are displayed below:

| BOOKS (in millions) | STUDENT ENROLLMENT (in thousands) | DEGREE | AREA (acres) |
|---|---|---|---|
| 4 | 5 | 3 | 20 |
| 5 | 8 | 3 | 40 |
| 10 | 40 | 3 | 100 |
| 1 | 4 | 2 | 50 |
| .5 | 2 | 1 | 300 |
| 2 | 8 | 1 | 400 |
| 7 | 30 | 3 | 40 |
| 4 | 20 | 2 | 200 |
| 1 | 10 | 2 | 5 |
| 1 | 12 | 1 | 100 |

Using a forward stepwise regression, show how each of the 3 factors affects the number of volumes in a college library.

8-3 We want to predict a student's success in college by a battery of tests. Graduating seniors volunteer to take our test battery and their final grade point average is recorded. Using a MAX-$R_2$ technique, develop a prediction equation for final grade point average using the test battery results. The data are as follows:

| GPA | HS GPA | COLLEGE BOARD | IQ TEST |
|-----|--------|---------------|---------|
| 3.9 | 3.8 | 680 | 130 |
| 3.9 | 3.9 | 720 | 110 |
| 3.8 | 3.8 | 650 | 120 |
| 3.1 | 3.5 | 620 | 125 |
| 2.9 | 2.7 | 480 | 110 |
| 2.7 | 2.5 | 440 | 100 |
| 2.2 | 2.5 | 500 | 115 |
| 2.1 | 1.9 | 380 | 105 |
| 1.9 | 2.2 | 380 | 110 |
| 1.4 | 2.4 | 400 | 110 |

8-4 Take a sample of 25 people and record their height, waist measurement, the length of their right leg, the length of their arm, and their weight. Write a SAS program to create a SAS data set of these data and compute a correlation matrix of these variables. Next, run a stepwise multiple regression, using weight as the dependent variable and the other variables as independent.

8-5 What's wrong with this program?

```
1       DATA MULTREG;
2       INPUT HEIGHT WAIST LEG ARM WEIGHT;
3       CARDS;
        (data lines)
4       PROC CORR;
5          VAR HEIGHT -- WEIGHT;
6       PROC STEPWISE;
7          MODEL WEIGHT = HEIGHT WAIST LEG ARM;
```

# Problem Solutions

## ANSWERS TO ODD-NUMBERED PROBLEMS

1-1 (a)
```
DATA COLLEGE;
INPUT ID AGE SEX $ GPA CSCORE;
CARDS;
1 18 M 3.7 650
2 18 F 2.0 490
3 19 F 3.3 580
4 23 M 2.8 530
5 21 M 3.5 640
```

(b)
```
PROC MEANS;
   VAR GPA CSCORE;
```

(c) Between the "INPUT" and "CARDS" lines insert

```
INDEX = GPA + 3*CSCORE/500;
```

Add to the end of the program

```
PROC SORT;
   BY INDEX;
PROC PRINT;
   TITLE 'STUDENTS IN INDEX ORDER'; (optional)
   ID ID;
   VAR GPA CSCORE INDEX;
```

1-3
```
1.    DATA MISTAKE;
2.    INPUT ID 1-3 TOWN 4-6 REGION 7-9 YEAR 11-12 BUDGET 12-14
3.         VOTER TURNOUT 16-20
(data cards go here)
4.    PROC MEANS;
5.       VAR ID REGION VOTER TURNOUT;
6.       N,STD,MEAN;
```

Line 3 - Variable name cannot contain a blank.
         Variable name too long. (Actually, if we
         had two variables, VOTER and TURNOUT the
         above INPUT statement would work, since we
         can combine LIST input with column
         specifications. However, for this problem,
         we intended VOTER TURNOUT to represent a
         single variable.)
         Semicolon missing after TURNOUT 16-20.
Line 5 - We probably don't want the mean ID. Also,
         would be more meaningful to use PROC FREQ
         for a categorical variable such as REGION.
Line 6 - Options for PROC MEANS go on the PROC line
         between the word MEANS and the semicolon.
         The options must have a **space** between
         them, not a comma.

```
          PROC MEANS N MEAN STD;
             VAR ---- ;

1-5       DATA PROB15;
          INPUT ID RACE $ SBP DBP HR;
          CARDS;
          (data go here)
          PROC SORT;
             BY SBP;
          PROC PRINT;
             TITLE 'RACE AND HEMODYNAMIC VARIABLES';
             ID ID;
             VAR RACE SBP DBP;


2-1   PROC FREQ;
          TABLES SEX;

2-3   (a)   PROC CHART;
               VBAR GROUP;

      (b)   PROC PLOT;
               PLOT Y*X;

      (c)   PROC SORT;
               BY GROUP;
            PROC PLOT;
               BY GROUP;
               PLOT Y*X;
```

Don't forget that you must have your data set sorted by the BY variables before you can use a BY statement in a PROC.

```
2-5   1   DATA;
      2   INPUT AGE STATUS PROGNOSIS DOCTOR SEX STATUS2
      3         STATUS3;
      4   (data cards)
      5   PROC CHART BY SEX;
      6      VBAR STATUS
      7      VBAR PROGNOSIS;
      8   PROC PLOT;
      9      DOCTOR BY PROGNOSIS;
```

Line 2 - PROGNOSIS has 9 letters.
line 2 - Not really an error, but it would be
         better to list SEX with the other
         demographic variables.
Line 2 - Again, not an error, but an ID variable is
         desirable.

Lines 2 and 3 - Boy, we're picky. If you have
STATUS2 and STATUS3, STATUS should be
STATUS1.
Line 5 - Two things wrong here: One, If you use a
BY variable, the data set must be sorted
in order of the BY variable; two, a semi-
colon is missing between PROC CHART and BY
SEX.
Line 7 - In case you thought this was an error, it
isn't. You **can** have two (or more) VBAR
statements with one PROC CHART.
Line 9 - A plot request is of the form Y*X not Y
BY X.

2-7 The most efficient program to read this data set and
compute means would be:

```
DATA PROB27;
INPUT ID TYPE $ SCORE;
CARDS;
1 A 44
1 B  9
1 C 203
  etc.
PROC SORT;
   BY TYPE;
PROC MEANS;
   BY TYPE;
   VAR SCORE;
```

Remember to sort your data set first, before using
a BY variable.

```
3-1   PROC FORMAT;
        VALUE FGROUP 1='CONTROL' 2='DRUG A'
                     3='DRUG B';
```

3-3 Between the DATA statement and the CARDS statement
insert

Method 1

```
IF 0 LE WEIGHT LT 101 THEN WTGRP=1;
IF 101 LE WEIGHT LT 151 THEN WTGRP=2;
IF 151 LE WEIGHT LE 200 THEN WTGRP=3;
IF WEIGHT GT 200 THEN WTGRP=4;
IF 0 LE HEIGHT LE 70 THEN HTGRP=1;
IF HEIGHT GT 70 THEN HTGRP=2;
```

(Note: You may use <= instead of LE, < instead of
LT, and > instead of GT.)

```
Then add      PROC FREQ;
                TABLES WTGRP*HTGRP;
```

Method 2

```
    PROC FORMAT;
        VALUE  WTFMT 0-100=1 101-150=2 151-200=3 201-HIGH=4;
        VALUE HTFMT 0-70=1 71-HIGH=2;
```

(Insert the following format statement before the CARDS statement.)

```
    FORMAT WEIGHT WTFMT. HEIGHT HTFMT.;
```

Then add      PROC FREQ;
                TABLES WEIGHT*HEIGHT;

3-5    Line 3 - The formats cannot be assigned to variables before they have been defined. Therefore, move lines 5 through 8 to the beginning of the program (before line 1).

Line 10 - PROC FREQ uses the keyword TABLES not VAR to specify a list of variables.

Line 10 - You cannot use the CHISQ option unless a two-way table (or higher order) is specified. That is, we could have written

```
    PROC FREQ;
        TABLES SEX*RACE / CHISQ;
```

Line 12 - You cannot use a BY statement unless the data set has been sorted first by the same variable.

```
4-1  (a)   DATA PROB41;
           INPUT X Y Z;
           CARDS;
           1 3 15          x vs. y  r= .965   p=.0078
           7 13 7          x vs. z  r=-.975   p=.0047
           8 12 5
           3 4 14
           4 7 10
           PROC CORR;
              VAR X;       (Note: SAS/PC vers 6.02 does not
              WITH Y Z;    have a WITH statement. Use
                           VAR X Y Z;)
```

```
      (b)  PROC CORR;          y vs. z  r=-.963   p=.0084
           VAR X Y Z;


4-3  (a)  PROC REG;           int. = .781 prob > |T|=.5753
           MODEL Y = X;  slope=1.524 prob > |T|=.0078

4-5  (a)  PROC PLOT;
           PLOT Y*X;

     (b)  PROC REG;
           MODEL Y = X;
           OUTPUT OUT=REGOUT PREDICTED=PY;
          PROC PLOT;
           PLOT Y*X PY*X='P' / OVERLAY;
```

4-7   Line 1 - Incorrect data set name, cannot contain a dash.

Lines 3-5 - These lines will recode **missing values** to 1, which we probably do not want to do. the correct form of these statements is

```
    IF X LE 0 AND X NE . THEN X=1;
```

                    or

```
    IF -999999 LE X LE 0 THEN X=1;
```

Line 10 - The options PEARSON and SPEARMAN do not follow a slash. The line should read

```
    PROC CORR PEARSON SPEARMAN;
```

Line 11 - The correct form for a list of variables where the "root" is not the same is

```
    VAR X--LOGZ;
```

Remember, the single dash is used for a list of variables such as ABC1-ABC25.

5-1   DATA HEADACHE;
      INPUT TREAT $ TIME @@;
      CARDS;
      A 40 A 42 A 48 A 35 A 62 A 35
      T 35 T 37 T 42 T 22 T 38 T 29
      PROC TTEST;
          CLASS TREAT;
          VAR TIME;

      Not significant at the .05 level (t=1.93, p=.083).

5-3 Use a paired t-test. We have

      DATA PAIR;
      INPUT SUBJ A_TIME T_TIME;
      DIFF = T_TIME - A_TIME;
      CARDS;
      1 20 18
      2 40 36
      3 30 30
      4 45 46
      5 19 15                            T=-3.00  p=.0199
      6 27 22
      7 32 29
      8 26 25
      PROC MEANS N MEAN STD STDERR T PRT MAXDEC=3;
          VAR DIFF;

5-5   Line 2 - Variable name HEARTRATE too long.
      Line 10 - Correct procedure name is TTEST.


6-1   DATA BRANDTST;
      DO BRAND='A','N','T';
          DO SUBJ=1 TO 8;
              INPUT TIME @;
              OUTPUT;
              END;
          END;
      CARDS;
      8 10 9 11 10 10 8 12
      4 7 5 5 6 7 6 4
      12 8 10 10 11 9 9 12
      PROC ANOVA;
          CLASSES BRAND;
          MEANS BRAND / DUNCAN;
          MODEL TIME = BRAND;

      F=4.91, p=.018. N is significantly lower than
either T or A (p <.05). T and A are not significantly
different (p > .05).

6-3 (a)
```
        DATA SODA;
        INPUT BRAND $ AGEGRP RATING;
        CARDS;
        C 1 7
        C 1 6
        C 1 6
        C 1 5
        C 1 6
        P 1 9
        P 1 8
        P 1 9
        P 1 9
        P 1 9
        P 1 8
        C 2 9
        C 2 8
        C 2 8
        C 2 9
        C 2 7
        C 2 8
        C 2 8
        P 2 6
        P 2 7
        P 2 6
        P 2 6
        P 2 5
        PROC GLM;
            TITLE 'TWO-WAY UNBALANCED ANOVA';
            CLASSES BRAND AGEGRP;
            MODEL RATING = BRAND|RATING;
            MEANS BRAND|AGEGRP;
```

   (b) Use the values from the MEANS statement in a) to plot the interaction graph.

   (c)
```
        PROC SORT;
            BY AGEGRP;
        PROC TTEST;
            BY AGEGRP;
            CLASS BRAND;
            VAR RATING;
```

6-5 Line 4 - Since this is a two-way **unbalanced** design, PROC GLM should be used instead of PROC ANOVA.

7-1
```
    DATA SHIRT;
    INPUT JUDGE 1 BRAND 2 COLOR 3 WORK 4 OVERALL 5;
    INDEX = (3*OVERALL + 2*WORK + COLOR)/6.0;
    CARDS;
    (data cards go here)
```

```
        PROC ANOVA;
           CLASSES JUDGE BRAND;
           MODEL COLOR WORK OVERALL INDEX = JUDGE BRAND;
           MEANS BRAND / DUNCAN;

7-3    PROC FORMAT;
           VALUE CITY 1='NEW YORK' 2='NEW ORLEANS'
                      3='CHICAGO' 4='DENVER';
        DATA PROB73;
        INPUT JUDGE 1-3 @;
           DO CITY=1 TO 4;
              INPUT TASTE 1. @;
              OUTPUT;
              END;
           END;
        FORMAT CITY CITY.;
        CARDS;
        0018685
        0025654
        0037464
        0047573
```

7-5  The DO loops are in the wrong order and the  OUTPUT statement is missing. Lines 2 through 8 should read:

```
        DO SUBJ=1 TO 3;
           DO GROUP='CONTROL','DRUG';
              DO TIME='BEFORE','AFTER';
              INPUT SCORE @;
              OUTPUT;
              END;
           END;
        END;
```

There are no other errors.


8-1  DATA TOMATO;
```
        DO LIGHT=1 TO 3;
           DO WATER=1 TO 2;
              DO I=1 TO 3;
                 INPUT YIELD @;
                 OUTPUT;
                 END;
              END;
           END;
        CARDS;
        12 9 8 13 15 14 16 14 12 20 16 16 18 25 20 25 27 29
        PROC REG;
           MODEL YIELD = LIGHT WATER;
```

8-3 
```
DATA PROB83;
INPUT GPA HS_GPA BOARD IQ;
CARDS;
3.9    3.8        680        130
3.9    3.9        720        110
3.8    3.8        650        120
3.1    3.5        620        125
2.9    2.7        480        110
2.7    2.5        440        100
2.2    2.5        500        115
2.1    1.9        380        105
1.9    2.2        380        110
1.4    2.4        400        110
PROC STEPWISE;
   MODEL GPA = HS_GPA BOARD IQ / MAXR;
```

8-5 Ha! No errors here. As a matter of fact, you can use this program for Problem 8-4.

# Index

**I**

ID statement, 14
Include command, 209
INPUT statement, 7,11
INPUT, column form, 7,40-41
INPUT, list, 11-12,40
IF statement, 49-50
INFILE statement, 207-208
Interaction, 123-129,149,152
Interactive, 4
Intercept, 76-77

**J**

JCL, 19-20
JOB statement, 5,20
Justification, 8

**K**

Kurtosis, 29

**L**

LABEL statement, 42
LAST., 67-68,231
Least significant difference, 114
Least squares, 75
Left justified, 8
LEVELS=, 27
LIBNAME, 209-210
Linear regression, 74-79
Lists of variables, 41,217-218
Log transformation, 88
Longitudinal data, 62-69
LSD, 114

**M**

Manuals, reference SAS, 2
Martians, 4
Mann-Whitney U-test, 99-102
MAXDEC=n, 10,40
Mean, 17,29
Mean, built-in function, 64,219
Mean-square, 78-79,91
Median, 29

Median test, 100-102
MERGE statement, 229
Missing value, 40,50
Mode, 29
MODEL, with ANOVA, 112
MODEL, with PROC REG, 75-76
Month format, 63
Month/day/year, 63
Multiple cards per subject, 63
Multiple comparisons, 114
Multiple regression, 178-195

**N**

Nested DO loops, 173
Nesting, in ANOVA, 147
Nonexperimental regression, 178,183-195
Nonparametric tests, 98-102
NOPRINT option, 65
Normal distribution, 18,29-30
Null hypothesis, 92

**O**

Observation, 6,69
One-tailed test, 103
One-way analysis of variance. See Analysis of variance
Options, 22
  with PROC MEANS, 22-23
  with PROC UNIVARIATE, 29-30
ORDER=, with PROC FREQ, 67-68
Ordinal scales, 99
Orthogonal designs, 118
OVERLAY option, 81
OUTPUT statement, 65,145
  with PROC MEANS, 65,220
  with PROC REG, 79-84
  with PROC SUMMARY, 221

**P**

Paired t-test, 103-106
Pearson correlation, 70-74
Pedhazur, Elazar J., 3
PLOT statement, 33
Pointer, 219
Post hoc tests, 114-117